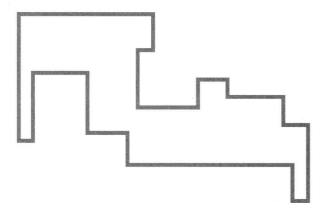

Managing the Construction Process

Estimating, Scheduling, and Project Control

Frederick E. Gould
PE, AIC
Wentworth Institute of Technology

Prentice Hall
Upper Saddle River, New Jersey   *Columbus, Ohio*

Library of Congress Cataloging-in-Publication Data

Gould, Frederick E.
    Managing the construction process : estimating, scheduling, and project control / Frederick E. Gould.
       p. cm.
    Includes index.
    ISBN 0-13-352337-3 (hardcover)
    1. Building—superintendence. I. Title.
TH438.G625 1997
690′.068—DC20

96-21797
CIP

Editor: Ed Francis
Production Editor: Rex Davidson
Design Coordinator: Julia Zonneveld Van Hook
Text Designer: STELLARViSIONs
Cover Designer: Proof Positive/Farrowlyne Associates
Production Manager: Pamela D. Bennett
Marketing Manager: Danny Hoyt
Electronic Text Management: Marilyn Wilson Phelps, Matthew Williams, Karen L. Bretz, Tracey Ward
Illustrations: Kurt Wendling

This book was set in Swiss 721 by Prentice Hall and was printed and bound by Quebecor Printing/Book Press. The cover was printed by Phoenix Color Corp.

© 1997 by Prentice-Hall, Inc.
Simon & Schuster/A Viacom Company
Upper Saddle River, New Jersey 07458

Printed in the United States of America

10 9 8 7 6 5 4 3 2 1

ISBN: 0-13–352337-3

Prentice-Hall International (UK) Limited, *London*
Prentice-Hall of Australia Pty. Limited, *Sydney*
Prentice-Hall of Canada, Inc., *Toronto*
Prentice-Hall Hispanoamericana, S. A., *Mexico*
Prentice-Hall of India Private Limited, *New Delhi*
Prentice-Hall of Japan, Inc., *Tokyo*
Simon & Schuster Asia Pte. Ltd., *Singapore*
Editora Prentice-Hall do Brasil, Ltda., *Rio de Janeiro*

With love to Nancy, Mckenzie, and Elliott

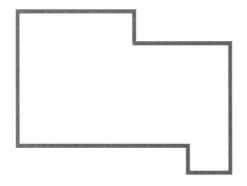

Preface

With the growth of the construction management profession, and the resulting expansion of construction management courses both in the major and as electives in many architectural and engineering school curriculums, a need has arisen to combine the various facets of the profession into one comprehensive text. Current texts either concentrate on a single function—scheduling, estimating, or project control—or cover the profession in an overview of project management.

In addition, practitioners who need a primer on current industry practices will find this text to be a good reference. The text is organized to cover all areas of the construction management industry, with an emphasis on maintaining a balance between theory and practice. Each of the four sections is introduced with background theory and fundamentals, which is followed by practical applications, frequent illustrations, sidebars written by industry professionals, chapter review questions, and a project highlight. The latter is a recurring feature, relating the teaching of each of the four sections of the book to an actual construction project, the recent renovation of two buildings at MIT. Through use of this text, students and practitioners—designers, owners, contractors, and construction managers—alike will gain knowledge of the building industry and the technical skills required to manage a construction project.

Section One, Construction Project Management, provides an overview of the industry. The roles and responsibilities of construction participants, organization of the project team, and factors affecting the project scope and timeline are all discussed in this section. Team play and concepts such as bonding, value engineering, and partnering are defined and emphasized. This section prepares the reader for a more thorough study of the major topics that follow: estimating, scheduling, and project control.

Section Two, Estimating, reviews the techniques and methods used in preparing the costs for a construction project. It discusses the quantity takeoff process, the establishment of unit prices, and the adjustment of costs for time and location. The section covers estimating in the context of a project's evolution. It demonstrates that as project information becomes better, the estimate becomes increasingly detailed, thereby feeding information back into the project to support sounder design and construction decisions.

Section Three, Scheduling, addresses the value of schedules and provides examples of different scheduling methods. Network based Critical Path Method is covered in the most detail. Activity definition, the creation of a logic diagram, the calculation of activity durations, and network calculations are all explained. Computer applications and examples of computer output are included.

Section Four, Project Control, concludes the book. This part examines how the estimate and the schedule are used to provide timely information to the owner and other project participants. In the preconstruction stage, this information can be integrated into a work plan that accurately projects resource usage. In the construction phase, this work plan allows the comparison of actual production to planned and provides feedback to the project team. This section looks at how the integration of the schedule and the estimate forms this work plan. Examples of integrated reports and a sidebar on Computer Integrated Construction complete this section.

Both the organization and the content of this book have been designed to allow it to serve as a useful reference for the practitioner as well as the student. In the classroom, the book will serve well as a teaching tool for the architectural, construction, or civil engineering student. The text provides an overview of all aspects of construction management, with enough practical examples that the student gets a real view into the world of construction management. As a reference for the professional, the book is organized so as to allow quick and easy access to information on current tools and practices of the profession—hence its utility to learners and experienced professionals alike.

Acknowledgments

Many people helped contribute to the writing of this book. I would like to acknowledge their efforts now. I particularly wish to recognize the contributions of Nancy Joyce, Senior Program Manager, Beacon Construction, who authored the project highlights and who served as technical consultant for the entire manuscript. Don Farrell volunteered the efforts of his construction photography firm, Farrell Associates, to provide the photographs used throughout the text.

I thank the following for reviewing the manuscript: James A. Adrian, Bradley University; Jeff Burnett, Washington State University; Charles Richard Cole, Southern College of Technology; Ellery C. Green, University of Arizona; and John Warsowick, Northern Virginia Community College.

Thanks to the following sidebar contributors for their real-world additions to the text: David Lash, Dave Lash and Company; Jeffrey Milo, Fischbach & Moore,

Inc.; Christopher Noble, Hill and Barlow; Kenneth Stowe, George B. H. Macomber Company; and Rory Woolsey, The Wool-Zee Company. I would also like to acknowledge the Wentworth Institute students for "agreeing" to be class tested on much of the book's content and, in particular, Matthew Viviano, who produced most of the Primavera plots used in the text.

The R. S. Means Company, especially their Engineering Department headed by Patricia Jackson, was a huge help in furnishing much of the cost data used in the estimating examples in the text.

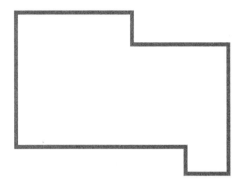

Contents

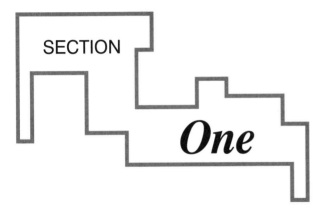

SECTION

One

Construction Project Management

Today's construction projects are managed by a team of people representing the owner, the designer, and the construction professions. These disciplines come together in many different ways depending upon the project type, the owner's sophistication, and the owner's time and budget concerns. Section One will look at the roles and responsibilities of the different disciplines. It will also examine different types of projects and the different ways that the professions work together to successfully manage a construction project. A project will be followed from an owner's idea, through design, and then through construction and project closeout. The activities that must occur at each point as the project moves through its life and the participants who must accomplish them will be discussed.

The construction management process will be described using terms and expressions unique to construction. As new terms are introduced they will be boldfaced and then defined in the text. Sidebars will also be used to provide longer definitions or related examples. An actual project will be introduced in this section and will be followed throughout the book to illustrate the complicated management of today's construction process. This project writeup is called a Project Highlight and will be found at the end of each section. The topics of estimating, scheduling, and control, which are the headings for the remaining three sections of this book, will be introduced in context in this section.

CHAPTER 1

Industry and the Project

STUDENT LEARNING OBJECTIVES ━━━━━━━━━━━━━━━━━

From studying this chapter, you will learn:

1. The nature of the construction industry.

2. The key elements that define a project.

3. The four major categories of construction projects.

4. The role of the project manager.

Introduction

The construction industry of today has been built on the needs of the world's inhabitants to provide shelter, conquer distances, harness energy, create public spaces, protect from natural disasters, and build historical monuments. These basic human needs have not changed over time even though the process and environment in which the "designer" and "constructor" operate have become increasingly more complicated. Rapidly escalating technology has made possible structures and processes unimaginable even to our grandparents. Construction projects such as the Skydome in Toronto, the "Chunnel" or Euro-Tunnel connecting France and England, super tall buildings, and maglev (magnetic levitation) rail systems are all projects that could only have been described as visionary as late as the 1960s (Fig. 1–1).

As design and material technologies have evolved, distances that can be spanned, heights that can be reached, and loads that can be carried have all increased. Cable-stayed bridges have made the 2-mile bridge a normality, every major city has a 500-foot skyscraper, and the high-speed "bullet train" has become more common. Specialists, people that focus on a singular aspect of a project, have evolved to make such feats possible. People are now able to make a living as an acoustical engineer, a CAD operator, or a scheduler and bring to the project specialized knowledge in one very focused area. The development of these specialists has created immense teams spanning many companies, states, and countries. As an example, today a Toronto based company might build a skyscraper in Dallas, which might be designed by a New York based architectural firm and be constructed by a Los Angeles based contractor using steel produced in Japan. The days of the master builder are no longer.

The computer has served as an important tool to help control this increasingly complex process. Today's construction project is designed on a computer using **CAD** (computer aided design); it is estimated, scheduled, and controlled by the contractor utilizing one of many different "off the shelf" software packages. Even this process is being optimized further by electronically linking the design and construction process. Today's designers and builders can preview tomorrow's construction operations on a computer; an owner can be given a tour through a building as it is being

Figure 1–1
Modern day super tall building.
Photo by Don Farrell

designed; and "object linking" allows a designer to compare the cost and constructability of competing components before deciding which to specify.

The environment in which the project is being built is also evolving. The laws, regulations, permits, and procedures that the project team must navigate to bring a project to completion is immense. A 1993 study conducted by the Electric Power Research Institute documented over 120 different environmental laws now in place as compared to only 13 in 1955 (Fig. 1–2).

In many cases, the sophistication of the city, state, or country in which the project is sited dictates the cost, schedule, and probability of successfully completing the project. This makes getting the approval to build often more difficult than the designing and construction of the project.

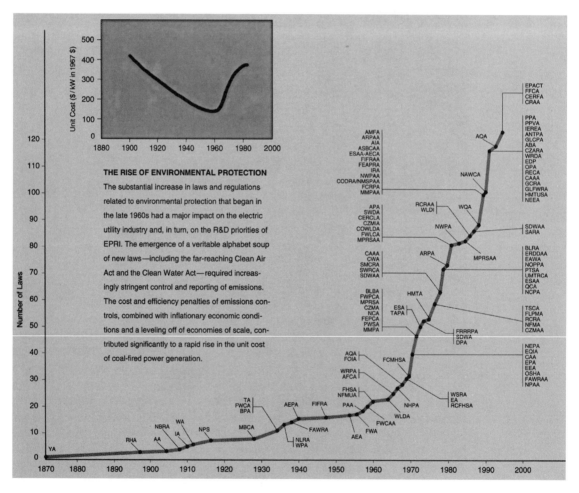

Figure 1–2
The rise of environmental laws impacting construction overtime.
Source: *EPRI Journal*, January/February 1993.

This chapter will look first at the construction industry in general, focusing on the characteristics and nature of the industry, followed by a look at its future. Next the construction project will be examined. What defines a project and how is it organized? An important factor in a project's organization is the person who leads the effort: the project manager. Personal qualities necessary to be a successful project manager will be discussed. The technical training required to be a successful architect or engineer (the usual background of a project manager) is so rigorous that little time is given to studying the leadership and management skills required to manage a complicated project.

The Construction Industry

When one analyzes the construction industry it is difficult to know where to draw boundaries, as so many people, organizations, agencies, and governments are affected by construction successes and failures. Construction activity (you often hear the term "housing starts") is often used as an indicator of the health and direction of the U.S. economy. This is because when construction activity is strong, many more people are employed.

As an example, consider a major interstate highway project (Fig. 1–3) and the number of companies and agencies that might be involved:

Federal:	Federal Highway Administration
	U.S. Secretary of Transportation
State:	Department of Environmental Protection
	Department of Transportation
	State Inspector General
Local:	Building Department
	Fire and police
Citizens:	Conservation Law Foundation
	Historic Review Commission
Business:	Business Roundtable
Designers:	Civil engineers
	Transportation engineers
	Architects
	Graphic designers
Construction:	Project managers
	Construction contractors
	Material suppliers
	Equipment suppliers
	Labor
Consultants:	Testing laboratories
	Surveyors
	Public relations

Figure 1–3
Central artery and tunnel project, Boston, MA.
Photo by Don Farrell

A major difference between the construction industry and other industries is that on a major project such as the one mentioned above, many businesses and agencies of varying size all come together for this one project. They will work together for a few years, then go on to another project with another group of participants. Construction is best described as a fragmented industry since these people and companies owe their allegiance more to their craft then they do to project leaders.

Construction is also more a service industry than a manufacturing or product based industry. Even though large products often are constructed, a project's success is more dependent on the people involved than on a particular piece of equipment, a process, or a patent. A project that can muster well-organized, skilled, and motivated people, with an effective communication system in place stands a good chance of succeeding. For this reason many public and private owners are focusing on team building and partnering sessions to establish strong leadership and communication systems. As seen in the Construction Team sidebar, written by Dave Lash, partnering has not yet reached its full potential.

Sidebar ▬▬▬▬▬▬▬▬▬▬▬▬▬▬▬▬▬▬▬▬▬▬

The Construction Team

> Fads will come and go. The fundamental fact of man's capacity to collaborate
> with his fellows in the face-to-face group will survive the fads and one day be
> recognized. Then, and only then, will management discover how seriously it has
> underestimated the true potential of its human resources.
>
> Douglas McGregor
> *The Human Side of Enterprise*

A recent industry trend is the convening of partnering retreats at the beginning
of projects. Usually called by the owner and guided by a professional facilitator,
these day-long conferences assemble members of a project team to discuss individual objectives, set common goals, and draft procedures for conflict resolu-
tion. Participants generally report improved understanding and communication
among team members, and fewer disputes.

Despite the introduction of these retreats, if I were entering the building
industry today, I'd want to know that teamwork remains the great unmet
promise of the industry. Perhaps partnering and teambuilding retreats signal the
dawn of a new era, but in the meantime, one would be wise to recognize the
four powerful winds that are continually blowing project teams off course.

1. The first wind we'll call *Ad Hoc*.   A project throws firms and individuals together to meet a specific, short-term need. Often strangers are expected to
mesh quickly, get the job done, and disband just as fast. Unfortunately, conflict
frequently arises because we lack the confidence and trust in others that comes
from sharing time and experiences together.

2. *Competing Interests* is the second wind.   Contract language shapes the
roles and responsibilities of the parties in profound and often adversarial ways.
Team members rarely sit at the table with well-aligned interests. Inevitably, each
organization strives to maximize its own position while minimizing its risk, often
with unintended and negative consequences.

3. *Culture* is the third wind.   Think of architects, engineers, and builders
as coming from different countries and you start to get the picture: each has its
unique temperament, language, history, rituals, and values. Throughout the
industry, misunderstanding is a daily occurrence.

4. The fourth wind is *Ignorance*.   The secrets of collaboration can be
learned—but they are rarely taught in the design and construction disciplines. For
decades, textbooks on project management ignored the vital human aspects of the
industry. And though some consciousness raising has begun, the initiator of most
partnering retreats is still the owner rather than the disciplines themselves.

My career, like that of almost everyone in the industry, has suffered the effects of these four winds. Despite a strong personal commitment to teamwork, and despite being part of over 40 project teams in the past 20 years, the most successful collaborations I've experienced in my life have been theater productions rather than building projects. Staging summer stock musicals, we endured prolonged tension, moments of extreme anxiety, and outbursts of anger. But the worst of times really can bring out the best in people. Despite turmoil, those actors, stagehands, designers, and technicians fought for a common goal with the understanding that they had only one simple choice: either succeed by trusting and supporting each other . . . or fail. In each instance, we chose to succeed, and by doing so we became great teams.

The word "team" itself gets us in trouble. If we had 50 words for "team" the way Eskimos have for "snow," perhaps we would learn to distinguish gradations of effective team behavior. I, for one, no longer think of the average project team as a "team" at all. Crick and Watson, who collaborated for years on the search for the structure of DNA, they were a team. NASA, when they put a man on the moon in less than a decade, they were a team. The engineers, poets, graphic artists, and marketers who built the Macintosh computer, they were a team. But a group of people representing different disciplines and companies, brought together on a short-term basis to design and build a single project, is not, in my thinking, a team at all. Instead, they are only a group of people given an opportunity, through honesty, desire, technique, and hard work, to forge a team and, by doing so, to share in one of life's most gratifying experiences.

Dave Lash
former Vice President
Beacon Construction Company
Boston, Massachusetts

The construction industry is composed of many small businesses due to its service nature, since it does not require a patent or a large capital investment to get started. Anyone with motivation and technical skill can start a business and be successful. However, because the industry is so closely tied to the cyclical nature of the economy, many small businesses are forced out during hard times.

Small construction businesses and the fragmented nature of the construction industry also affect the amount of investment made in new technologies. A 1986 *Engineering News-Record* article on the competitiveness of the U.S. construction industry showed investment in research and development (R&D) as a percent of sales compared to other industries as follows:

U.S. Pharmaceuticals   7%
Aerospace/Automobile   4%

Japanese Construction   3%
U.S. Construction   .01%

The nature of the industry tends to discourage investment in research since most projects are pushed through with tight time tables. New ideas always carry the risk of failure and a lawsuit. If a company invests in a successful new idea or technology, a competitor will often use this idea on its next project. This occurs since many construction industry ideas and technologies are not patentable and therefore are available for anyone to use.

The Players

The Owner

The owner, also called the client, is the person or organization that will pay the bills as well as receive the ultimate benefits of the finished project. The owner is responsible for determining what the project will include (also called the **scope** of the project), when the project can begin and must end (the schedule), and how much he or she can afford to spend (the budget). In most cases the owner relies upon the advice of other people to establish these project parameters. Large companies or institutions that are involved in constructing major facilities have entire divisions set up to handle this process. Facility engineers, facility managers, and planners are a few of the job titles for people who specialize in this sort of work. Small businesses or companies that do not do a lot of construction may rely upon outside consultants to assist them through the process. Project managers, construction managers, and design professionals can provide this service.

Owner organizations can be broken into two major categories, public and private. A **public** agency exists for the ultimate benefit of the citizenry, the general public. Since the project is paid for from public funds, statutes exist which describe how the project is to proceed. Examples of public projects would be a town library, an interstate highway, or an army barracks. Statutes require that these projects be publicly advertised with all qualified and responsible bidders given the opportunity to compete. Wage rates and bonding requirements are also commonly stipulated.

Private organizations can be described as any individual, partnership, corporation, or institution that builds a project for its own use or for resale. A private organization has much more freedom as to how it proceeds with a project. Private organizations often invite selected designers and builders to compete for their projects. Examples of private projects would be individual homes, shopping malls, or some hospitals and universities.

In order to achieve success on a project, owners need to define quickly and accurately the project's objective. They need to establish a reasonable and balanced scope, budget, and schedule. They need to select qualified designers, consultants, and contractors to work on their project and they need to put in place an effective control system to stay informed about the project. As will be described throughout the course of this book, the project type, the organizations involved, and contract methods chosen all dictate different levels of owner involvement and control.

The Design Professional

Examples of design professionals are architects, engineers, and design consultants. The major role of the design professional is to interpret or assist the owner in developing the project's scope, budget, and schedule and to prepare construction documents that will be used by the construction contractor to build the project.

Depending on the size and sophistication of the owner, the design professional can be a part of the owner's facilities group or an independent, hired for the project. In some cases the design professional and construction contractor together form a design-build company.

In almost all cases the design professional is a licensed, registered professional who is responsible for the physical integrity of the project. In the United States each state licenses architects and engineers who are allowed to practice within that state. To attain a license the professional must demonstrate competency by a combination of degree(s), experience, and examination. Many states have reciprocal arrangements with other states, so that the license of one state is accepted in another.

The purpose of registration is to guarantee the public that the construction project will be built in accordance with acceptable design standards and will comply with local safety standards. Professional associations, material suppliers and manufacturers, testing associations, and building code agencies are all producers of standards

TABLE 21
Recommended Uniform Roof Live Loads for APA RATED SHEATHING(c) and APA RATED STURD-I-FLOOR With Long Dimension Perpendicular to Supports(e)

Panel Span Rating	Minimum Panel Thickness (in.)	Maximum Span (in.)		Allowable Live Loads (psf)(d)							
		With Edge Support(a)	Without Edge Support	Spacing of Supports Center-to-Center (in.)							
				12	16	20	24	32	40	48	60
APA RATED SHEATHING(c)											
12/0	5/16	12	12	30							
16/0	5/16	16	16	70	30						
20/0	5/16	20	20	120	50	30					
24/0	3/8	24	20(b)	190	100	60	30				
24/16	7/16	24	24	190	100	65	40				
32/16	15/32	32	28	325	180	120	70	30			
40/20	19/32	40	32	—	305	205	130	60	30		
48/24	23/32	48	36	—	—	280	175	95	45	35	
60/32	7/8	60	48	—	—	—	305	165	100	70	35
APA RATED STURD-I-FLOOR(f)											
16 oc	19/32	24	24	185	100	65	40				
20 oc	19/32	32	32	270	150	100	60	30			
24 oc	23/32	48	36	—	240	160	100	50	30	25	
32 oc	7/8	48	40	—	—	295	185	100	60	40	
48 oc	1-3/32	60	48	—	—	—	290	160	100	65	40

(a) Tongue-and-groove edges, panel edge clips (one midway between each support, except two equally spaced between supports 48 inches on center), lumber blocking, or other. For low slope roofs, see Table 22.

(b) 24 inches for 15/32-inch and 1/2-inch panels.

(c) Includes APA RATED SHEATHING/CEILING DECK.

(d) 10 psf dead load assumed.

(e) Applies to panels 24 inches or wider applied over two or more spans.

(f) Also applies to C-C Plugged grade plywood.

Note: Shaded support spacings meet Code Plus recommendations.

Figure 1–4
Example of a design/construction guide.
Courtesy of the Engineered Wood Association

that must be understood and conformed to (see Fig. 1–4). Stamping a drawing certifies that the design professional is now responsible for the integrity of the design.

The Construction Professional

The responsibility for the interpretation of the contract documents and the physical construction of the project rests with the construction contractor. In a traditional arrangement where the owner, design professional, and contractor are separate companies, the contractor would be termed a **prime contractor** and would be contractually responsible for delivering a completed project in accordance with the contract documents. In most cases the prime contractor divides the work among many specialty contractors called **subcontractors**. On a large project these subcontractors may also divide up the work into even smaller work packages (see Fig. 1–5).

As projects continue to get larger and more technical and as owners seek earlier and more accurate pricing and scheduling, many variations have begun to occur in the contractor's role. Contractors are getting hired earlier in the process and are being asked to provide technical, cost, scheduling, and constructability advice to the owner and the design professional. In this shift into preconstruction involvement, the contractor acts as a construction consultant. Taking this in a Pure Construction Management arrangement the construction professional, owner, and design professional

Figure 1–5
Contractor hierarchy.

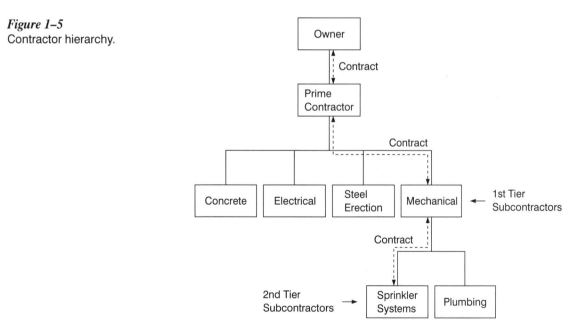

Owner initiates contract with prime contractor.
Prime contractor signs contracts with 1st tier subcontractors.
1st tier subcontractors may sign contracts with 2nd tier subcontractors.

work as a team through the whole project directly managing the subcontractors without the need of a prime contractor. All of these different methods of arranging the professionals is explained in detail later in this book.

Future Outlook

Looking to the future, the U.S. construction industry faces many challenges. Our economy has become global with opportunities for work throughout the world, while at the same time U.S. companies continue to experience more foreign competition in our own backyard.

Robots and other high tech equipment are already replacing workers in occupations that are dangerous or undesirable. This trend will continue to evolve, eroding more and more unskilled positions, requiring those who enter the work force of the

Figure 1–6
Working in a constrained environment, "cutting and patching to match," and material disposal make renovation projects a particular challenge.
Courtesy of New England Conservatory & Walsh Brothers, Inc.: The Jordan Hall Restoration
Photo by Don Farrell

future to be more highly skilled. This necessity will put increased pressure on unions and our educational system to provide that technical training.

Government regulations, environmental permits, and other bureaucratic controls continue to grow. Projects will also continue to get larger and more technical, requiring more specialized people, high tech equipment, and better control systems. This trend will require that tomorrow's project leaders have technical, business, organizational, and leadership savvy to successfully complete their projects.

Related to these needs is the fact that more and more of tomorrow's projects will be built in congested locations on existing sites. These "renovation-type" projects create very tricky issues dealing with disposal of waste, asbestos, PCBs, and the like (see Fig. 1–6).

The project may also uncover historic artifacts or burial grounds. As the project is constructed, building occupants need access, and the roadways, power supply, and so forth must be kept operational, which requires detailed advanced planning that might involve utilizing temporary structures and facilities.

The traditional challenge to bring a project in on time, under budget, and of the highest quality will continue to exist, but methods of accomplishing all this will be

Figure 1–6, *continued*

examined and perfected. Owners will push the "design-builders" to provide earlier and more precise budgeting, tighter and faster schedules, and more value for the money. These demands will be made in the face of potentially rising prices and predictable shortages of key materials. High speed information processing with computers linking owner, designer, builder, and supplier will improve communications and speed the delivery of the project.

The key survival skill in the construction industry of the future will be knowledge and awareness of the state of the industry and where it is going. The survivors and future leaders will be those who have the technical skills, business skills, and leadership skills and the ability to work with people. Being able to adapt, knowing where to find the answer and how to write a good memo, and having the ability to sway a hostile audience to your side are the kinds of skills that will deliver the project of the future.

The Construction Project

A project is defined, whether it be in construction or not, by the following characteristics:

1. A defined goal or objective
2. Specific tasks not routinely performed
3. A defined beginning and end
4. Defined deliverables
5. Resources being consumed

The goal of a construction project is simple—it is to build something. What differentiates the construction industry from other industries is that its projects are large, built on-site (the factory is brought to the job site), and generally unique. There's only one Golden Gate Bridge, Sears Tower, and Alaskan Pipeline.

Every project can be clearly broken down into a series of logical, definable steps which will become the roadmap for the project. The project team will start at the beginning of the list, and when they get to the end the project is over. Projects are characterized as having a single starting and ending point with all the work in the middle. The uniqueness of each project characterizes the high risk nature of project management. Because projects are generally one-time ventures, a bad roadmap can lead the team in the wrong direction, wasting time and money. Project starts and finishes are negotiable and totally dependent on the work that the owner decides to do and the money available.

The defined deliverables are what the owner establishes as the program and what is further clarified in the contract documents by the design professional. A certain quantity of work will be completed in accordance to certain specifications within a certain time frame. It is imperative that the project team clearly delineate the desired performance and think about how this performance will be measured and controlled.

Construction projects consume tremendous quantities of resources, all of which are paid for by the owner. Time, money, labor, equipment, and materials are all examples of the kinds of resources that are managed and controlled by the project team. Efficiently managed projects minimize, balance, and forecast resource consumption for the owner.

Projects begin with a stated purpose, a goal established by the owner which is to be accomplished by the project team. As the team begins to design, estimate, and plan out the project, the members learn much more about the project than was known when the goal was first established. This often leads to a redefinition of the stated project goals.

The project management process has been described as an iterative process: the project team establishes a goal with the understanding that more will be learned about the project, which then redefines the goal. This process will be repeated over and over again (see Fig. 1–7).

The process has also been characterized as a spiral: the team begins the project knowing very little about it, but as time moves on more and more is learned about the project as to goals, available technologies, and requirements, forcing a redefinition of the project as the spiral broadens (see Fig. 1–8).

What both of the last two figures illustrate is that projects begin with estimates and best guesses as to what the end result of the project will be once completed. Owners have to begin investing in a project with estimates, often purchasing land and hiring financial, marketing, and legal advice. They need to begin designing the project and to pay for estimating, scheduling, and project management support. They are often forced to make commitments on completion dates to end users. This illustrates the importance of these initial estimates and why it makes sense for own-

Figure 1–7
A project as an iterative process.

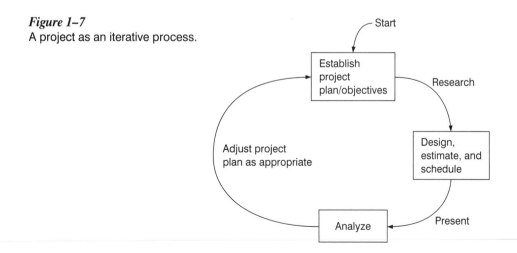

Figure 1–8
Project information spiral.

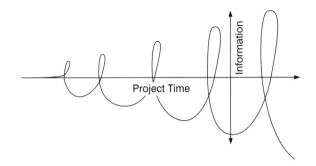

ers to get the best advice possible at the earliest possible time in the life of the project (see Fig. 1–9).

Figure 1–9 visually shows the importance of getting good advice early in the project. As can be seen, the ability to influence decisions falls off sharply as time on the project passes. This indicates that early decisions have much greater importance than later ones. For instance, an early decision such as deciding whether to cross the harbor using a sunken tube tunnel or a suspension bridge is more critical than the decision of which project sitework contractor to select. The figure also illustrates that project costs start slowly, but increase sharply once the project enters the construction phase. Clearly the most money is spent for construction materials and labor. Taking the analysis one step further, upfront advice is cheap as a percent of the total project cost; even though hiring one more consultant or designer may be expensive, if that one can help steer the project in the right direction it will be money well spent.

People and organizations move in and out of the project, so the actual team managing the project continuously changes. The goal of the project, as stated earlier, evolves, so the people whose business it is to direct the project have to very carefully ensure that all involved are working from the same plan. The minutes of meetings, project update meetings, published budgets and schedules, conference calls, and updated organizational charts are all examples of the tools necessary to keep a project on track.

Figure 1–9
Level of influence.

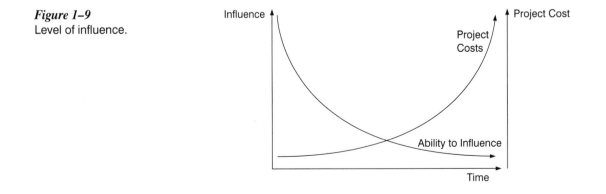

Figure 1–10
S curve of projects

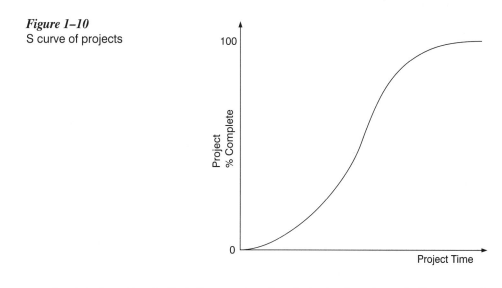

A related matter is that the construction industry is extremely fragmented, bringing people from many different companies to the project. Companies that come together on the project may enter it with different organizational goals, which create conflict if not properly managed. An example is the coordination between a material supplier, who needs to balance the production demands thereby delaying a delivery to your project and a subcontractor who needs to wrap up work on your project to get to another commitment.

Projects build up slowly as workers and equipment are brought to the project and mobilized. Early on only a few activities may occur, but once mobilization is complete work proceeds at a rapid pace until the end, when production slows down and the last remaining items are wrapped up. A cumulative production curve, also called an **S curve** (Figure 1–10) reflects this. S curves can be used to represent both production quantities and cash expenditures to date.

Categories of Construction Projects

Most designers and builders tend to focus their efforts within specialty areas, focusing on particular types of projects. Four informal categories of construction are:

Residential
Building construction
Infrastructure and heavy construction
Industrial

These distinct types have evolved because major differences exist in the way the projects are funded, in the technologies involved, and in the manner in which the designers, builders and owners interact.

Residential

Residential construction projects include the construction of individual homes as well as small condominium and apartment building complexes. These projects tend to be privately funded by individual owners for their own use or by developers for profit. Such projects are typically designed by architects, but in some cases a single home may be designed by the individual home owner or builder. There has been some movement towards standardization with companies offering premanufactured homes and products, but in general most homes are still built on site; they are often called "stick built."

This industry tends towards the use of fairly low technologies and requires little investment to enter. For these reasons the industry is characterized by large numbers of small designers and builders and suppliers. When the economy is strong these small companies do well, but when times are tough many of these companies go out of business. Interest rates and government policy towards housing investment also influence the health of this industry. About one-third of construction spending is on residential construction (see Fig. 1–11).

Figure 1–11
Residential construction.
Courtesy of Five Bridge Farm
Photo by Don Farrell

Building Construction

Examples of building construction projects would be office buildings, large apartment buildings, shopping malls, and theaters. In terms of annual construction expenditures this project category is the largest, although it is heavily dependent on the economy of a particular region. Houston in the 1970s and the Northeast and California in the 1980s all saw tremendous growth and construction success, but because of overbuilding these regions saw little construction in the years to follow as overbuilt real estate was absorbed.

Projects such as these are designed by architects with engineering support and are generally built by general contractors. Most of these projects are privately funded, though some projects like schools, courthouses, and city offices are publicly funded. The technical sophistication of building construction projects is greater than residential construction as is the investment necessary to enter. These factors provide for fewer players than in residential construction (see Fig. 1–12).

Figure 1–12
Commercial building construction.
Courtesy of New England Deaconess Hospital & Walsh Brothers, Inc.
Photo by Don Farrell

Infrastructure and Heavy Construction

Some examples of this category of construction are roadways, bridges, dams, and tunnels. These projects are designed principally by civil engineers and built by heavy construction contractors having engineering backgrounds or support. These projects are usually publicly funded and therefore are sensitive to governmental policy. At the time of this writing there is a tremendous focus on rebuilding the nation's infrastructure, with many large funded projects. These projects tend to be long in duration and thereby less sensitive to the ups and downs of the economy, and they involve the heavy use of equipment (see Fig. 1–13).

Figure 1–13
Heavy engineering construction.
Photo by Don Farrell

Industrial

Steel mills, petroleum refineries, chemical processing plants, and automobile production facilities all serve as examples of industrial facilities. These projects are defined more by the production activities within the facility than by the facility itself. The design and construction of the shell is dependent on the needs of the process and production equipment. In the United States most of these facilities are privately funded although in developing countries these facilities may be publicly funded. Quality and time are extremely important in these projects. Because these projects are the most technical of all, only a few designers and builders are qualified to bid on any one type of facility. Since the process technology is critical, the designers and builders need to work together throughout the project. Often the facility designer and constructor are the same company (see Fig. 1–14).

Figure 1–14
Industrial construction.
Courtesy of Mr. Raymond Bourque
Photo by Don Farrell

The Project Manager

Having reviewed the nature and key participants of the construction industry, as well as the characteristics of a project, it is now important to look at the role of project management and the people who do it.

The first step in the management process is to arrive at a clear definition of the goals of the project and to understand what problems the project is designed to solve. Depending on the project type, as just mentioned, the role of the owner, designer, and construction professional varies, as does the time and budget allowed. It is the job of these people, the project management team, to define the project and to arrive at a detailed description of the program that is acceptable to most. Additionally, a good project manager is a negotiator and needs to be able to keep the client happy as space or equipment needs are compromised for other project needs.

A second role of the project team is to investigate alternate solutions to the stated problem. The project team sometimes needs to do a tremendous amount of investigation to develop alternative solutions. The design professionals need to find available technologies, while the construction professionals investigate cost and schedule implications. The environment in which the project will be built needs to be continually studied, and the end users (if known) need to become involved in the process. Flexibility and oral and written communication skills all become important to the project manager. During this project stage the team is involved in problem solving. Many ideas and criticisms will materialize, and the project manager's job is to maintain open and constructive communication.

Evaluating the alternative solutions and arriving at a single program that incorporates the best value for the client is the next step. Value is an intangible that is a measure of quality balanced against budget and schedule. The client needs to define what its expectations are so that alternative solutions can be completely measured and evaluated. To work through this step the project manager must quantify and process available information and must be able to evaluate technical alternatives and suggest refinements. He or she must know how to read technical drawings as well as be able to get along with the project team members. The successful project team fosters cooperation and allows all involved to learn from each other.

The next step is to develop a detailed plan to make the selected program a reality. This work plan lays out for all involved specific tasks and responsibilities and establishes for each step all resource, budget, and schedule parameters. It should be developed involving all the key project team members—designer, owner, construction professionals, and consultants. These players should be involved in establishing the key targets for the project as well as in measuring progress and making adjustments as the project proceeds. While working at this stage, project managers must be able to estimate costs, develop schedules, and prepare budgets for the entire project.

Next, the project team needs to implement the plan and control the project. As the plan moves to the field the project team needs to mobilize all contractors, suppliers, and vendors who will be involved on the project, and then evaluate their progress and make adjustments. To accomplish these tasks the project manager needs to be able to negotiate contracts, use computers, interview and evaluate subcontractors and vendors, coordinate employees, and evaluate progress using technical drawings.

As the project draws to an end the project team needs to make all necessary final adjustments and evaluate the final performance of the project. Individual project participants also must look toward their next project involvement. The project manager must be able to balance the short-term project goals with long-term career goals and look toward increasing project knowledge and advancement within the company. As the job comes to an end the project manager must be able to assess the success of the project. If the project did not end up as planned, all must ask themselves why not? It is important to learn from this work so that the same mistakes are not repeated on future projects.

Conclusion

This chapter introduced the construction industry, its key players and major project types, the characteristics of a project, and the role of the project manager. The remaining chapters will build on the unique characteristics of the industry as well as the project environment and will further detail the steps that need to be taken to develop the program and—more specifically—to estimate, schedule, and control the construction project.

Chapter Review Questions

1. Research and development expenditures are higher in construction than in other industries.

   ___ T   ___ F

2. Public sector projects are those funded by tax dollars through cities, towns, states, and the federal government.

   ___ T   ___ F

3. Most construction subcontractors would be classified as licensed registered professionals.

   ___ T   ___ F

4. The amount of information that a project manager must consider increases as a project moves towards completion.

   ___ T   ___ F

5. Dams, bridges, and highways would be classified as commercial building projects.

   ___ T   ___ F

6. Site selection and financing would be the responsibility of which project team member?
   a. Owner
   b. Designer
   b. Construction project manager
   d. Trade subcontractor

7. This category of projects is often funded by public dollars and is termed "infrastructure."
   a. Residential
   b. Commercial building
   c. Heavy engineering
   d. Industrial

8. Which of the following statements would *not* be true with respect to the future of the design-build industry?
   a. The industry will become more global.
   b. There will be an increased demand for more highly trained workers.
   c. Permitting and regulations will decrease.
   d. Hazardous waste disposal and product recycling will become more of a factor.

9. As projects move on in time the ability to change the project becomes _____ difficult and _____ expensive.
   a. more—less
   b. less—less
   c. less—more
   d. more—more

10. Which of the following is *not* a characteristic of a project?
    a. Having a specific goal
    b. Having a defined beginning and end
    c. Resources being consumed
    d. Usually being performed only once
    e. Never being found outside the construction field

Exercises

1. Interview a local architect, engineer, owner representative, construction manager, or contractor. Determine that person's typical project responsibilities, concerns, and project goals. Present your findings to the class.

2. Photograph several local projects. Identify the designer, builder, and owner of the project. How long did the project take to be designed and built? What was the approximate cost of the project? Identify the project category. Present your findings to the class.

The Project Management Process

STUDENT LEARNING OBJECTIVES

From studying this chapter, you will learn:

1. The principal project phases.
2. The roles and responsibilities of each project team member during each project phase.
3. How to manage a project through the project's life cycle.

Introduction

Chapter 1 introduced the construction project, the industry, and the key players. It discussed key characteristics of a project and the individual skills that a good project manager possesses. This chapter will also examine the construction project, focusing on the many small steps that make up a project. The chapter will begin with the conceptual phase, then work through the design, procurement, construction, and close-out phases of a project. The chapter will continue to discuss the ingredients that make up a successful project and project manager. It will also identify individual responsibilities of the owner, designer, and construction professional. The vocabulary of the construction industry will begin to play a prominent part in this chapter since many of the activities that the project team must accomplish are not defined by common household expressions. Terms such as *bid package*, *prequalification*, *bonding*, and *punchlist* will be defined as they are introduced in the chapter.

The order in which activities occur in a construction project and the question of which professional accomplishes each of them depends on how the owner decides to manage the project. For example, the owner may decide to hire an architect to design a shopping mall and a contractor to build it after it has been totally designed. Because a total design has been completed, the owner can ask for a bid or fixed price before construction begins. This arrangement is called the traditional or conventional method of delivering a project (discussed further in Chapter 3). The advantage to an owner is that the price for the project is known before construction begins.

Another possibility might be for the owner to hire both an architect and a construction manager before design begins. The advantage to the owner of following this route is that by coordinating design with construction the actual building of the shopping mall can begin before the design is finished, saving project time and possibly money. This is called **phased** or **fast-tracked** construction. (See Fast Track sidebar.)

Sidebar

Fast-Track Construction

To successfully complete any project many tasks need to be accomplished by the project team. The owner must define the requirements, the designer needs to translate the requirements into contract documents, and the construction professionals need to organize and manage the physical construction in accordance with the contract documents.

The normal way to accomplish these tasks is to complete each phase, then move on to the next. Conceptual planning is first, then design, then procurement, and finally construction (see Figure A, top). Each project phase builds on the last; the owner communicates needs to the designer, and the designer clarifies those needs through contract documents to the builders. This approach is understood fairly well by the project participants and is logical, but compared to a fast-tracked project it is very time consuming.

Figure A
Comparison of the conventional, sequential arrangement to the fast-tracked approach.

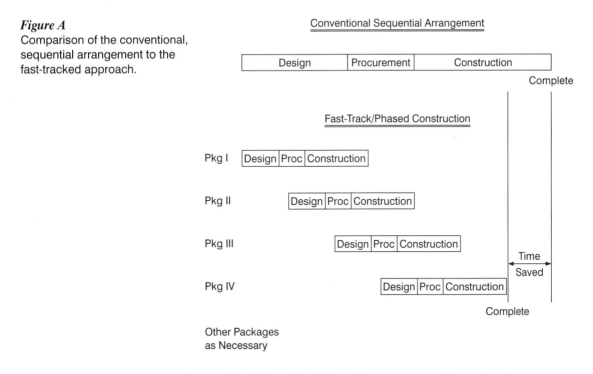

In a fast-tracked (also called phased) arrangement, the project is approached in less of a linear fashion (see Figure A, bottom). In a fast-tracked approach the project is broken down into smaller pieces (called work packages), with each package designed and constructed separately. By breaking the project down the work that can logically be done first (e.g., site clearing and excavation) can be designed and performed while later work (paving, roofing, finishes) is still being designed.

Fast-Track Challenges

A fast-tracked project can clearly save time and money for an owner, but this savings does not come without risk. The greatest risk stems from the fact that the owner must begin construction without a complete design and a detailed and complete estimate. If construction begins and design problems occur on the later stages of the project, forcing a budget increase, the owner is at that point committed to more than a simple redesign.

Good communication between the designer and construction manager is also essential since they and the owner are often working on concurrent tasks that must be coordinated. The designer's design schedule must be precisely tied to the bid and award of each work package and the detailed construction schedule.

Overall, the owner/designer/construction manager team must guard against rushing into construction too early; it is essential that the project be thoroughly investigated and that an accurate program, estimate, and schedule be developed and be completely in hand.

The advantage of fast-tracked projects is clearly speed. However, the coordination and cooperation between the owner, architect, and construction manager must be very good, otherwise owner costs can skyrocket. Also, because the design is not totally complete when the construction manager is hired it is difficult for the owner to know the real cost of the project before construction begins. Chapter 3 will focus on the different delivery methods that are used and the advantages and disadvantages of each.

The discussion of the activities through the following project phases will assume that the project is fast-tracked and that a construction manager and designer have been hired early in the preconstruction phase. The discussion will assume an active owner staff that will be working with the construction manager and the designer throughout the entire project.

Project Chronology

Preconstruction

The preconstruction phase of a project can be broken down further into conceptual planning, schematic design, design development, and contract documents. The bidding and award phase is also a preconstruction activity, but activities that occur in that stage will be discussed in the Procurement Phase section. The preconstruction phase requires continuous owner and designer involvement and interaction since during this project stage the ideas and requirements of the owner must be clearly translated into contract documents by the owner. The construction manager's role during this phase is to support the design process, judge the design for constructability, look for ways of reducing cost, prepare for construction, and provide competent cost and schedule information.

Conceptual Planning

The conceptual planning stage of the project is a busy and important time for the owner. During this phase of the project the owner makes decisions that set the tone for the project. During this stage the owner hires key consultants including the designer and construction manager, selects the project site, and establishes a conceptual estimate, schedule, and program. The most critical decision that is made during this project phase is whether to proceed with the project or not.

Early on in the conceptual stage the owner must work with the end users of the project to determine what functions the completed project will serve and to balance these with available funds and time. An owner that does a lot of building (e.g., McDonalds) knows very well what the building will contain, how much it will cost, and how long it will take to construct, but an owner building a new university library, a municipal building, or a new airport must spend a tremendous amount of resources to work through this stage of the project. This process is iterative, meaning it goes around and around. The organization puts together a list of what it wants; that gets priced and scheduled; then the list gets longer or shorter accordingly. In some projects, particularly renovations, it is very hard to determine exactly what will have to be done up front before the project proceeds. For example, in renovation work asbestos may be uncovered and have to be removed. In the airport project mentioned above a local neighborhood group may begin to fight the project and may make additional demands which must be funded.

Working through the issues just outlined is extremely difficult, but very important. To confuse the issue further the owner may have alternate sites on which this project may be built, each with different considerations. An insurance company planning to build an office building both for its own use and for rental purposes would have to look at the cost of the building site, the rent per square foot it could collect, the cost of hired labor, and the cost of construction in the area, to name but a few issues. All of this data would also have to be projected to the future, to accurately estimate costs when the project would begin to be built and used. This analysis would have to be redone for each site under consideration.

On major projects, particularly heavy construction and industrial facilities, the permitting and regulatory process can be both critical and very difficult to predict. The entire U.S. nuclear power industry has almost disappeared because of the difficulty and cost of acquiring permits and abiding by all the regulations. Even getting the permit to locate a department store can be very expensive and difficult—ask Wal-Mart! Owners must budget the cost and time of this process as accurately as possible.

To successfully work through this stage of the project the owner must get good advice as quickly as possible. The owner should select and hire the appropriate design and technical consultants to properly analyze, program, estimate, and schedule the project. The more technical and politically sensitive the project, the more sophisticated support is needed. Referring back to the Level of Influence Curve illustrated in Chapter 1, it can be shown that it makes sense to get as much good advice as possible early on during the time when the most critical decisions are made. Taking this one step further, it does not make sense to select your design and technical professionals based solely on fee, but take into account their experience and reputation. Remember that the consultant's advice is something the owner has to live with for the remainder of the project. As an example, consider what the costs to the owner would be if the foundations are designed incorrectly due to a mistake in the analysis of subsurface conditions of a site by a geotechnical consultant.

In sum, the conceptual planning stage of a project is when the owner needs to gather as quickly as possible as much reliable information as possible about a project. This process may require hiring quite a number of design and technical consultants to help if those resources are not available within the company. Once the information is formulated the owner needs to make a decision as to whether or not to proceed with the project (called a **go/no go decision**). If the decision is go, the owner needs to select a site, establish a program, a conceptual estimate, and a master schedule. The designer and construction manager should also be hired at this stage.

Schematic Design

The completion of the schematic design phase represents approximately 30 percent design completion for the project. During this phase of the project the design team investigates alternate design solutions and alternate materials and systems, as well as supporting the **value engineering** program (see Value Engineering sidebar). While the owner and designer are working out project requirements, the construction manager should begin to establish the **work package**, or **bid package**, format for the project (see sidebar on Work Packages).

The construction manager also needs to identify all the **long-lead items** for the project. A long-lead item could be a critical piece of equipment or material that takes a significant amount of time to manufacture and bring to the site. An example might be structural steel, elevators, compressors, or custom cabinets.

This is the stage of the project when the design and construction professional first come together. It is important that the two professionals establish a good working relationship for the duration of the project. As the designer continues with the design of the project, the construction manager needs to refine the estimate and schedule for the project. A decision by the construction manager to tighten the schedule impacts the designer just as specifying a specific material impacts the construction manager.

Sidebar

Value Engineering

Value engineering is a creative, organized approach the objective of which is to optimize the cost or performance of a project. To best apply value engineering, strong communication and cooperation must occur between all project team members throughout the entire design and construction process. The first documented use of value engineering was by the General Electric Company during World War II. GE applied value engineering as a strategy to identify and utilize substitute materials as appropriate due to shortages caused by the war.

The value engineering process can be broken down into five distinct stages:

1. Informative
2. Speculative
3. Analytical
4. Proposal
5. Final Report

Value engineering is best applied early in the designing/building process and should consider all aspects of a project which affect cost and provide value to an owner. The cost to maintain and operate the project over its lifetime (life-cycle costs) should also be considered in the analysis. As an example, a more expensive heating system may be justified since it will consume less fuel and cost less to maintain. Decisions as to system choice, operating and maintenance costs, and even aesthetic appeal need to be made with complete knowledge as to cost, schedule, quality, and material availability. The value engineering process is designed to assist the project team in making informed decisions about project design.

In the informative phase the project manager, owner, and designer need to alert all project personnel to be on the lookout for alternate materials and methods. Throughout the entire design and construction process, project personnel should be encouraged to offer suggestions which should be documented by the project manager. Questions such as: What is it?, What does it do?, How much does it cost?, and What amounts are used? should be asked by project personnel.

The speculative phase is in many ways a continuation of the informative phase. The questions noted above are still asked as well as questions such as: What is its function?, What is the value of the function?, and What else will perform that function?. This phase can be likened to a brainstorming session with every idea noted and no question considered "dumb." The purpose of this phase is to create ideas and begin to identify value to the owner, but to defer judgement until later.

The analytical phase is when the alternative ideas generated above are researched, estimated, and evaluated as to technical feasibility by the designer. Suggestions deemed feasible are evaluated as to whether they are better, equal, or inferior to the original design. This phase may involve outside consultants, vendors and suppliers as well as the owner, designer, and construction professional. The impact of alternative ideas on both the design and construction schedule is critical since the later a good idea is received, the greater the possibility that redesign or reconstruction may have to occur. The impact of schedule on costs must clearly be considered in the analysis.

After analysis, ideas which are deemed possible and which in the opinion of the project team generate cost savings or increase project value to the owner should be presented for final approval. This is the proposal phase of the project. The last phase, the Final Report, is when a final tally is made on actual costs saved. Contractors who are working under a formal value engineering incentive clause may earn additional money for ideas which they submit that are accepted by the owner. That makes the Final Report and the final costs shown in it important to both parties.

Value engineering is an important function for all project team members and should be both formalized and encouraged. Among some designers value engineering has a bad name. They see it as cost cutting without consideration given to the overall design of the project. That should not be the case, since a good value engineering program should continually involve the designer, who must always maintain the responsibility for the technical competency of the design. A good value engineering program should generate more and better ideas earlier in the process and ultimately should provide a better and more cost effective design for the client.

Value Engineering References

Brown, Robert J., PHD, and Rudolph R. Yanvek, PE. *Life Cycle Costing*. Atlanta, Georgia: The Fairmont Press, 1980.

Dell'Isola, Alphonse J. *Value Engineering in the Construction Industry*, 2d ed. New York: Construction Publishing Co., 1974.

Dell'Isola, Alphonse J., and S. J. Kirk. *Life Cycle Costing for Design Professionals*. New York: McGraw-Hill, 1981.

Miles, Lawrence D. *Techniques of Value Analysis and Engineering*, 2d ed. New York: McGraw-Hill, 1972.

O'Brien, James J., PE. *Value Analysis in Design and Construction*, New York, McGraw-Hill, 1976.

Sidebar ▬▬▬▬▬▬▬▬▬▬▬▬▬▬▬▬▬▬▬▬▬▬▬▬▬▬▬▬

Work Packages

The work package, also called a bid package, is the organizational tool used to "break down" the construction project. Most construction projects are broken down into smaller components that are organized by common work elements akin to the way the trade contractors in the region are organized. Work packages are also used to control the timing of a project. This is done through the use of a master schedule. The work package defines an element of work which can be priced and scheduled, allowing the project team to coordinate and budget complicated construction projects.

The designers and construction managers need to decide early on what work will be performed in each of the work packages. Every work element must be included. Care must be taken to avoid omitting any work or describing the same work element in two different work packages. This is a difficult process since many work elements can be accomplished by various work packages, and some work needs to be closely coordinated between two different trade contractors. An example of this might be the controls for the heating, ventilating, and air conditioning systems (HVAC). The controls, thermostats, and the like need to be used in the balancing of the mechanical systems, but they also require electrical work. Are they put into the mechanical or the electrical work package? The approach that needs to be taken in such a situation is to follow common practices of the area, and in union work to abide by local negotiated trade agreements.

The work package contains all the information necessary to describe the work that needs to be performed: it should include drawings, specifications and addenda, and the conditions under which the work must be performed; a contract to be signed if the bid is accepted; general and special conditions; and the construction schedule. If a bid breakdown is required or the owner is furnishing certain items, this also needs to be specified. An invitation to bid invites the interested contractor to submit a bid at a certain time and place, and when the bidder does submit a bid the company will usually use a bid form which will identify a price, overhead and profit for markups on changes, and a statement that the bidder agrees to complete the contract if the bid is accepted and will provide required bonds.

Preparation of the bid packages needs to involve all of the project team. The designers are primarily responsible for preparing the technical requirements as described in the drawings, specifications, and addenda, but clearly must break up the work so as to describe only the work included per each bid package. The construction manager also must generate contractor interest, determine common work practices in the area, and communicate this to the designer. Both the designer and the construction manager must work from the overall master schedule to prioritize their work.

The fewer the bid packages, the easier the coordination becomes between the management team and the trade contractors. However, as the packages get fewer in number, they get bigger, and it gets more difficult to generate interest among smaller trade contractors. Less interest leads to higher bid prices. Bigger work packages also become less manageable and tougher to phase, tending to lengthen the overall construction time.

Particularly if the project is going to be fast-tracked (i.e., construction started before the design is totally complete), it is important that the bid package format as well as the design, bid and award, and construction schedule be established at this point (see Fig. 2–1).

As the designers proceed toward completing the construction documents, the construction manager must continue to establish an estimate for the work as well as identify acceptable construction subcontractors. The format of the estimate, the subcontractors considered, as well as the administrative and control procedures established must be coordinated between the designer and the construction manager.

Figure 2–1
Master schedule.

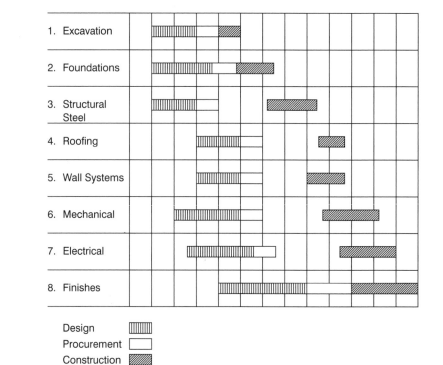

Design, procurement and construction phases are overlapped by work packages.

A value engineering program should also be established and presented to all project participants during this stage of the project. The designer's role will be to evaluate the technical merit and acceptability of the alternative ideas; the construction manager's task will be to estimate the cost and schedule impact of the alternatives and document the cost saving ideas. It is the job of every project team member to look for alternate solutions.

The schematic design stage ends with a design, estimate, and schedule presentation to the owner. This presentation may very well include several different design solutions with accompanying estimates and schedules. It is the job of every team member to ensure that participants are heading towards the same project goal. This is a formal review point and a good time to raise questions and concerns about the different designs being presented. In most cases only one scheme will be carried forward from this point, so it is important that either the best design or the best ideas from each design be selected. Many good—and sometimes conflicting—ideas that affect design, program, estimate, and schedule will and should be raised, but as the formal schematic review ends and the project moves into design development it is essential that a single vision be accepted and followed by all.

Design Development

The design development phase takes the project from about 30 percent to approximately 60 percent design. It is the time when the design team will be evaluating and selecting all of the major systems and components of the project. During this phase of the project the design team is involved in the evaluation and finalization of all the architectural components and project systems. The project's budget and schedule continue to be monitored and adjusted. The construction manager is involved in evaluating potential contractors and in preparing to secure the necessary permits. Good communication between owner, designer, and construction manager is critical during this phase because system choice dictates use, appearance, construction, and operating costs. Examples of system choices that might have to be made in a commercial building project are:

Structural system:	Concrete vs. structural steel
Mechanical system:	Multizone vs. single zone
	Gas vs. electric heat pump
Elevator system:	Two large vs. three small
Curtain wall:	Precast concrete panels vs. glass
Roofing:	Built-up vs. single-ply
Foundation:	Caissons vs. piles

The design team's role will be to recommend each of the project's systems; the owner's role will be to verify that each recommended system is acceptable to the end user; and the construction manager's job is to evaluate the cost and schedule implications of each selection. The team should be looking for a balance between all of the systems that are chosen. As an example, each of the systems can be priced and calculated as a percent of the total cost of the project. This percentage can then be compared to past similar projects. This process allows the owner to see how this project compares to past projects and also to see which components of the project are costing high and low respectfully. Project components that are high without adequate justification become good candidates for cost savings. On large projects this process becomes even more important since architects and engineers design and specify systems often independently of each other and need to get together periodically to coordinate and compare their design approaches and to check costs.

The continued estimating and scheduling of the project through the design development phase is critical to the success of the project since it allows the project managers to detect early on a design in excess of budget and to ensure schedule continuity. If an adjustment needs to be made it can happen early and be coordinated with all the design participants. This is preferable to a forced redesign when bids are received from the contractors and they all exceed the budget. That situation forces the project to be redesigned and then rebid, increasing both the cost and duration of the project. If the cost of the project needs to be reduced, the owner is informed early on and is able to make an educated decision as to what to change in the program. As an example, an owner may elect to choose a lower quality system, reducing construction cost at the expense of operating cost. Another owner may elect to reduce the size of a project while maintaining quality. These decisions require active owner involvement and good design and construction advice.

Concurrent with the design decision-making process is the preparation for construction by the construction manager. All the necessary permits and information required needs to be identified and organized. Applications must be prepared, hearings scheduled, and reports prepared and submitted, and all these efforts need to be coordinated with the design process.

It is important at this time to begin to generate contractor interest in the project as well as to find out about the overall market conditions of the area. Some of this may have been done during the site selection process, but now all information should be reevaluated and used in the preparation of the current estimates. It is to the project's benefit to get active contractor involvement in the project since this will lead to good competition and better prices. In highly technical projects, key subcontractors may be hired now to critique the design and to offer advice about materials and methods. The **prequalification** of bidders is important at this point, in preparation for the procurement phase of the project. To prequalify bidders is to evaluate their capability to successfully complete the work assigned to them in a timely manner and with high quality. The contractors' financial condition, past project performance, employees, equipment owned, and other current projects are all looked at as part of the prequalification process. Ideally, the construction manager will find five to ten interested and qualified contractors for each planned bid package.

As the design development phase of the project ends, the design of the project should be about 60 percent complete and should include good system definition. Most of the project should be quantified and the actual unit prices developed. A detailed network schedule should be in progress and the necessary permits identified and scheduled. All key subcontractors should be identified and involved in the estimating, scheduling, and design process as necessary.

Contract Documents

The contract documents phase of a project involves the final preparation of the documents necessary to define each of the bid packages (see again the Work Packages sidebar). This work would include the preparation of the technical documents, drawings, and specifications, as well as the general conditions and all other necessary supporting documentation. This process can be greatly complicated when the project is being fast-tracked since the construction of the project will have begun while the construction documents for the later parts of the project are still being prepared. Clearly, as work moves from design into construction it becomes more difficult and expensive to make changes.

The designer's major task during this project phase is to prepare all of the documentation which will define the technical requirements of the project. This includes primarily the working drawings and the technical specifications. When the project is to be phased, this work becomes more difficult since the contract documents for each part of the project need to be precisely coordinated with the other packages to avoid overlap or omission (see Fig. 2–2). This breakdown needs to be closely coordinated with the construction manager, who is finalizing contractor bidders for the different bid packages.

The master schedule for the project becomes important during this phase since the designer's completion of the technical requirements for each bid package needs to be coordinated with the advertising and bidding out of the work packages to the qualified contractors, followed by the start of construction. Ideally, the completion of

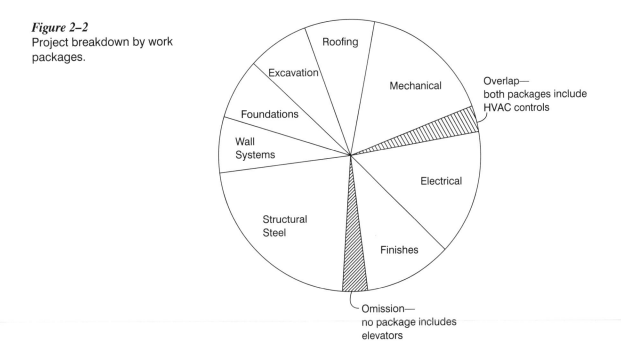

Figure 2–2
Project breakdown by work packages.

design, the advertising and selection of the contractor, and the beginning of construction proceed smoothly with a manageable overlap. (See again the Fast Track sidebar.)

Prior to going out to bid, the contract documents need to be closely reviewed by the construction manager and appropriate owner personnel. This is the last time that any discrepancies can be corrected before the work will be viewed and priced by the contractors, so it should be a very thorough review. Particular attention should be paid to the coordination between the different work packages as well as to what items are being furnished by the owner and the construction manager. If long-lead items have been purchased by the owner, it is imperative that this be stipulated in the appropriate bid packages.

With the contract documents now nearing completion, the construction manager is able to prepare a complete and detailed estimate for the project. This is sometimes called a **fair cost**, or owner estimate, which defines what the owner organization (construction manager) sees as the fair value of the work which will be competitively priced by identified, prequalified bidders. This detailed estimate, which will be explained more precisely in Part Two of this book, is important to the project manager for several reasons:

1.   It defines the "fair cost" for the work, allowing the owner to verify contractor pricing as well as negotiate with the trade contractors.

2.   Putting together this estimate is an excellent way for the owner/construction manager team to familiarize themselves with the project in preparation for construction. It also provides a final review of the documents, allowing final adjustments or clarifications to be made to all bidders in the event of mistakes. Notices of changes in the contract documents during the bid are called **addenda**.

3.   The estimate also provides the owner with final, accurate cost information which, when integrated with the finalized network schedule, projects the owner's cash needs on a day-to-day basis. This is called a **cash flow analysis** and will be covered in Part Four of this book.

As the project moves toward construction, the construction manager needs to prepare for management of the field operations. A field staff needs to be assigned, organized, and housed as necessary, possibly requiring the renting of office space and trailers. Utilities, signage, fencing, police details, and all other necessities will have to be contracted by the owner or construction manager. Before any construction can begin, a building permit must be obtained.

This phase completes the design work for the project, which, as explained above, can be done on a bid package by bid package basis to support phased construction, or can be done in total with a single contractor bidding on the entire project. The differences between these approaches will be explained in more detail in the next chapter. With a building permit in hand the project moves into the construction phase, as contractors bid and are selected by the owner or construction manager. A final estimate and schedule and an integrated cash flow analysis are completed and are now used to manage the remainder of the project.

Procurement Phase

The procurement phase, which is also called the **bidding and award** phase, is the time when the project formally transitions from design into construction. As was explained in discussing the contract document stage, this phase can overlap the design stage if the project is fast-tracked. In a traditional arrangement without phased construction, the bidding and award of the contract will occur only once, as a single general contractor is selected. If a construction manager is hired and the project is phased, each work package will be advertised and bid out individually.

This phase of a project should not be rushed since the contractors that are selected are directly responsible for the interpretation and completion of the work. If the contractors are properly prequalified and are given adequate time to prepare accurate prices for the project, the chances are better that the work will be done well. Conversely, if the contractors are randomly selected and given little time to prepare the bids, mistakes will be made and the construction phase will be difficult to manage. Contractors who mistakenly underbid will be less cooperative and will be constantly looking to rectify their mistake. Disputes are also likely to occur over contract interpretation, which may result in increased legal costs. In the worst case scenario, overextended contractors will not be there when needed, holding up other contracts, and if pushed may go out of business. All of the above can be avoided by properly prequalifying the trade contractors who bid the job. Contractors can be required to submit a bond which will provide some insurance to the owner, but prequalification should still occur (see Bonding sidebar).

Sidebar ▬▬▬▬▬▬▬▬▬▬▬▬▬▬▬▬▬▬▬▬▬▬▬▬▬▬▬▬▬▬

Bonding

What Is a Bond?

A contractor who submits a bond to an owner is providing to the owner a guarantee that in the event he or she does not perform as stated, the bonding company, an independent third party, will cover the owner's damages to the amount of the bond. For a contractor to receive a bond, the contractor's ability to perform the work must be independently evaluated by the bonding company. The type of project, the company's resources, and the size of the project are a few of the matters considered before a bond is provided. This process is in many ways similar to the prequalification process that the construction manager performs. The process of screening the contractor is essential to the bonding company (also called a surety) since the bonding company is obligated to pay if the contractor defaults.

The cost of a bond is factored based on the type of bond provided, the face value of the bond, and the risk rating of the contractor. The higher the risk, the greater the price. Since the cost of the bond is passed on to the owner, an owner may not ask for a bond in certain cases. A bond may not be required on a project where the contractor has done considerable work for the owner in the past.

Types of Bonds

The three most common types of bonds are the bid, the performance, and the payment bond. The bid bond will be discussed first. When an owner receives bids on a project, the owner may ask that along with the bid a bid bond be included. The bid bond's purpose is to guarantee that if the owner accepts the contractor's bid, the contractor will enter into a contract with the owner in accordance with the contract terms. If the contractor does not, the owner has the right—depending on the conditions of the bond—either to collect the value of the bond or to require the bonding company to pay the difference between the submitted bid and the next lowest bid.

A second type of bond is called a performance bond. A performance bond guarantees that the contractor will complete the project in accordance with the contract provisions. If the contractor defaults on the project, the owner can turn to the bonding company to complete the project up to the face value of the bond. Most bonds are submitted at 100 percent of the contract amount, but on some occasions they may be less.

The third type of bond that may be required is called a payment bond. A payment bond stipulates that the contractor, when paid by the owner, will then pay its suppliers and subcontractors. If they were not paid, the subcontractors and suppliers would be forced to file a lien on the property. A lien would encumber the title, possibly forcing a foreclosure sale to pay off the project debts. Without a bond, the owner would be forced to pay twice for the work not paid off by the contractor. A payment bond would force the surety to pay the contractor's debts, avoiding the liens.

Conclusion

A bond by a surety company does not totally prevent all risks to an owner, but it does serve to help minimize the damage caused by poor contractor performance or nonperformance. On large projects general contractors may also require bonds from their subcontractors to help protect their interests. Bonding does not substitute for prequalification since the need to use a bonding company to complete the work will cause tremendous disruption with corresponding losses to the owner.

The procurement phase begins with a public advertisement which will notify all interested bidders, or an invitation to bid which will notify specific bidders. If contractors have already been identified and prequalified, those contractors will be invited to bid. In the case of public work all interested and qualified bidders will be invited to submit a bid (see Fig. 2–3).

This invitation notifies potential bidders about the work, as well as instructing bidders as to where to pick up or review the contract documents and when and where the bids are due. If bonding is a requirement, that will also be mentioned. On some projects a prebid conference will follow, which will allow interested bidders to tour the project site and ask questions of the owner, designer, and construction manager.

As bids are received, the construction manager along with the designer needs to carefully review all of them. It is not unusual for bids to be received which suggest alternate methods and materials as part of the contractor's price. Alternate ideas should be encouraged, as they form an important component of a comprehensive value engineering program. The review process must be done thoroughly and must be fair to all bidders. This becomes difficult when bidders begin to suggest alternate ideas that may exceed or may not meet the project specifications. If a contractor suggests an alternative that exceeds the job requirements and is still low bidder, that firm should get the job. However, if a contractor suggests an alternative that exceeds the contract requirements, but is not the low bid, then those bidders in contention should be allowed to rebid the job. Good ideas that are cost savers but which do not meet the contract requirements should be negotiated with the low bidder. After bids are reviewed, tabulated, and adjusted, the construction manager will recommend a contractor and a contract will then be signed between the contractor and owner.

As prices are received on the bid packages, the construction manager will be able to get a true sense of the "bidding climate" as compared to the fair cost estimate. These actual prices will then replace the budget prices, and as more bid packages are placed under contract a progressively more accurate budget will be created for the project. As the job moves under contract, the remaining risk for variance lies in unforseen conditions such as bad weather, strikes, differing site conditions, designer error, unbuildable details, dimensioning errors, or subcontractor nonperformance. As the procurement phase ends, the project should have transitioned into construction, with all labor and material contracted, and with both its schedule and estimate updated.

Construction Phase

The construction phase of the project is defined as the actual physical construction of the project. If the contract documents have been well prepared and reflect the owner's true goals and if the subcontractors that are selected are responsive, then the construction phase should go smoothly. This phase requires a tremendous amount of monitoring and control to support construction activities, which calls for the involvement of the entire project team.

The designer's role during the construction phase is first to review the contractor **shop drawings** and samples to validate conformance with the contract. This process is critically important because if unacceptable materials or methods are

DWSD Local Contracts
Section 00030 Advertisement

1. Bids — The City of Detroit, Michigan will receive sealed Bids for the Work delivered to the Detroit Finance Department, Purchasing Division, 912 City-County Building, Two Woodward Avenue, Detroit, Michigan 48226, until 2:00 P.M., local time, on December 12, 1995, when all Bids duly received will be opened publicly and read aloud.

2. Project description — The Work, Adams Road Station Improvements, Contract No. DWS-708, includes, but is not necessarily limited to construction of a 10 million gallon wire-wound, prestressed concrete reservoir, including related sitework; clearing, paving, grading, overflow detention pond and dewatering pump chambers, landscaping, yard piping, valve pits, an above-grade addition to the existing pump station for housing electrical equipment, including foundations, masonry walls, structural steel, heating and ventilation, lighting and power. In the pump station: installation of two (2) 18 mgd reservoir pumps with 1500 hp motors and related piping, valves, and fittings, extension of the existing reservoir fill line with related valving, and fittings including orifice plate, removal and replacement of the existing eddy current variable speed drive device on Line Pump L1 with a variable frequency drive system ventilation system improvements, power and instrumentation. Removal and replacement of the existing house service transformer, and installation of fan cooling at the existing primary service transformer.

3. Bidding Documents — Beginning November 22, 1995 sets of Bidding Documents may be obtained from the Engineer at the DWSD Contracts Section, 1401 Water Board Building, 735 Randolph, Detroit, MI 48226 on Business Days between the hours of 8:00 A.M. and 4:30 P.M. Copies may be obtained upon payment of $100.00 per set, which will not be refunded.

Bidding Documents will be shipped only if the requesting party assumes responsibility for all related charges.

4. Bid Security — Each Bid shall enclose Bid Security, as specified in the Instructions to Bidders (and Section 00310, if a Bid Bond), in the amount of five percent (5%) of the Bidder's Bid.

The site is located at the Northwest corner of Adams Road and I-75 in Bloomfield Township, MI, as shown on the Drawings.

5. Site Tour & Pre-Bid Conference — A site tour shall be conducted for prospective bidders, on November 27, 1995, during the hours of 10:00 a.m. and 3:00 p.m., local time. The site tour will be at the Detroit Water and Sewerage Department's Adam's Road Station, located at the Northwest corner of Adams Road and I-75 in Bloomfield Township, MI. A pre-bid conference will be held on November 28, 1995, at 10:00 a.m., local time in the 16th floor Board Room, Water Board Building, 735 Randolph, Detroit, MI 48226.

Prospective Bidders and others interested in the Work are encouraged to attend the pre-bid conference and tour. For questions concerning the specification, Drawings or site tour contact Mohamad Jaber at (313) 224-5763. For information on the pre-bid conference, call Jacque-

lyn Jordan at (313) 224-4707. Addenda may be issued in response to issues raised at the pre-bid conference and tour, or as the Owner and/or Engineer may otherwise consider necessary.

6. Prequalification — All tank contractors are required to be prequalified. The bidder is required to state on the face of his sealed proposal the name of the prequalified tank contractor. Sealed proposals which do not state the name of a prequalified tank contractor will be returned to the bidder unopened.

Natgun Corporation, Wakefield, Massachusetts and Preload, Inc., Garden City, New York are prequalified for precast, prestressed concrete tank construction. Additional tank contractors, if any, seeking prequalification shall submit detail design drawings and calculations along with their record of previous experience in the design and construction of circular precast, prestressed concrete tanks and method of prestressing to the Engineer for review and approval no later than fifteen (15) days prior to the date set for receipt of bid. Within ten (10) days prior to the date of receiving bids, the engineer will publish a list of any additional prequalified tank contractors.

7. Local Contracting Requirements — Section 00300 contains Executive Order #4. Executive Order #4 requires that thirty percent (30%) of the total dollar value of all contracts let by the City be awarded to either Detroit-Based Business or Small Business Enterprises.

Executive Order #22, stipulates that worker hours for this Contract shall be performed by not less than 50% bona fide Detroit residents, not less than 25% minorities, and less than 5% women.

DWSD maintains a Voluntary goal of 25% Minority Business Enterprises and 10% Women Business Enterprise participation for this contract.

Section 00300 contains Executive Order #4.
Section 00460 contains Executive Order #22.
Section 00800 contains City Ordinance #20-93 which requires prevailing wages and fringe benefits rates for this project.

Each Bidder shall be required to provide necessary information to obtain a clearance from the Detroit Human Rights Department and Income and Property Tax clearance from the Detroit Finance Department.

All local requirements applicable to this contract are reproduced in their entirety in the Bidding Documents. Potential Bidders are specifically directed to review requirements before submitting their bid.

8. Contract Times — The Contract Times and the associated liquidated damages are specified in Article 3 of the Agreement.

9. Award — Subject to any agreed extension of the period for holding Bids, Bids shall be open for acceptance by the Owner for one hundred twenty (120) days after the date of Bid opening. In addition, the Owner expressly reserves the right to reject any or all Bids, waive any non-conformances, to issue post-Bid Addenda and re-Bid the Work without readvertising, to readvertise for Bids, to or withhold the award for any reason the Owner determines and/or to take any other appropriate action.

State of Missouri
Division of Design and Construction

Sealed Bids for Northeast Correctional Center Construction, Bowling Green, MO, Proj No. 30-936-95-0050(B) will be received at the Div of Design and Construction, Ofc of Adm, State of MO, RM 730, Truman State Ofc Bldg, PO Box 809, 301 W. High St, Jefferson City, MO 65102, until 1:30 pm, 01/18/96, and then publicly opened. A 5% bid security is required.

Pre-Bid Meeting: 01/03/96 - 10:00 am, Community Center, 201 W. Locust St., Bowling Green, MO Obtain plans and specs from Booker Associate, Inc., 1139 Olive St, St. Louis, MO, 314/421-1476. Upon payment of a refundable fee of $300 per set, by certified, cashier's or company check payable to Booker Assoc.

East Bay Municipal Utility District
Request for Qualifications
for Engineering, Geotechnical and
Environmental Services,
Raised Pardee Dam and Associated
Improvements

The District is preparing to contract for engineering, geotechnical and environmental services to assist with this project. Pardee Dam is a concrete gravity arch dam that impounds Pardee Reservoir on the Mokelumne River in Amador and Calaveras Counties in California. The enlargement of Pardee Reservoir by up to 200,000 acre feet is one of several projects the District is concurrently analyzing to meet its need for water during a drought. The District's Board of Directors will decide at a future date which project(s) to construct.

The District anticipates carrying out investigations, studies and design work for several facilities, including raising the main dam, modifying or replacing the spillway, modifying the powerhouse, raising or replacing a secondary dam near the Jackson Creek outlet, replacing the Highway 49 bridge over the Mokelumne River, raising and/or replacing the outlet tower and modifying aqueduct facilities, and modifying recreational facilities. Environmental analysis for this work may also be included in this contract or covered in a different contract. A separate RFQ for environmental services has been prepared and consultants are invited to respond to either or both of the RFQs.

Copies of the RFQs may be obtained through the District's Water Supply Improvement Division at 375 11th Street, Mailstop 305, Oakland, CA 94607-4240 by calling project secretary Ms. Ann Reis at (510) 287-1197 between 8:00 am and 5:00 pm PST or by faxing Mx. Reis at (510) 287-1295. The RFQ details consultant submission requirements. Consultant Statements of Qualifications (SOQ's) must be received by 5:00 pm PST, January 5, 1996. A selection committee will evaluate the SOQ's and the most qualified teams will be invited to submit formal proposals.

Figure 2–3

Bid announcements for two construction projects, as well as a request for qualifications for engineering, geotechnical, and environmental design services.

used, the integrity of the design may be compromised (see Hyatt sidebar). The thoroughness of and time taken for the review of the submittal, as illustrated in the sidebar, is critically important. The contract specifications will generally specify how long each agency involved has to conduct the review. This review time must be monitored closely. The scheduling section of this book illustrates the importance of review time. Material deliveries may be located on the critical path of the project such that every day that a submittal is late will hold up the project one day.

Another role of the designer during construction is to monitor the construction quality and to assist the construction manager in authorizing **progress payments** to contractors. It is normal for the trade contractors to request payment at the end of the month for the work completed during that period. Before authorizing owner payment, it is important to verify that the work has in fact been done and has been done correctly.

Contract changes, which are made to the contract documents after issuance, can be a major responsibility for the design professional during the construction phase. A change order can be issued for a number of reasons, which will be explained in greater detail in the next chapter.

As has already been discussed, one of the major goals of the preconstruction process is to avoid changes by constantly verifying and reverifying the scope and design of the project. The reason that this is so important is that implementing of changes during construction can be disruptive to the project. When changes are made, work may have to be stopped, work that has been completed may have to be demolished and redone, and the work schedule may have to be adjusted. All of these factors cost money and time because most change orders are negotiated and not competitively bid, so the owner must pay top dollar. The designer's role is to make the adjustments to the design and to implement the necessary change with the least impact to the ongoing construction. This is a process that should be performed in cooperation with the construction manager.

One of the first responsibilities of the construction manager is to support the **mobilization** efforts of all of the trade contractors on the job site. The trade contractors need to submit proof of insurance, a detailed schedule for their work, a safety program, a quality control program, an equal employment opportunity program, and a projected schedule of values (explained in Section Four), which will be used to govern progress payments.

As the work proceeds, the construction manager will be working with the designer to verify the quality and progress of work as well as to manage the change order process. A major task of the construction manager will be to document specific activities of each trade contractor and to maintain and update the network schedule. As the work of each trade nears completion, the construction manager will prepare a **punchlist** and coordinate necessary corrections. A punchlist is a list of all items of work that the owner/designer/construction manager require be completed before they will release final payment. The trade contractor's list may be quite different, so it is up to the construction manager to negotiate the difference.

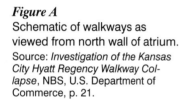

Figure A
Schematic of walkways as
viewed from north wall of atrium.

Source: *Investigation of the Kansas
City Hyatt Regency Walkway Col-
lapse*, NBS, U.S. Department of
Commerce, p. 21.

Sidebar

Hyatt Regency Walkway Collapse

One hundred thirteen people were killed and 186 injured when two suspended walkways collapsed in the Hyatt Regency Hotel in Kansas City on July 17, 1981. This was the most devastating structural collapse in the history of the United States. This is an accident that could have been prevented if a better coordinated engineering review had taken place in the shop drawing process.

The hotel's design called for three walkways to span the atrium at the second, third, and fourth floors. The original design specified six single 46-foot rods to run from the ceiling through the fourth floor box beams and on through the second floor box beams. The box beams were made up of a pair of 8-inch channels with the flanges welded toe to toe such that the weight of the platforms was carried on washers and nuts attached to the hanger rods. The third floor walkway was offset and supported independently on its own set of hanger rods (see Fig. A).

During the course of construction, shop drawings were prepared by the steel fabricator which suggested that a set of two hanger rods replace the single hanger rod on the second and fourth floor walkways. Thus, a rod would extend from the roof framing to the fourth floor, and a second rod would run from the fourth floor walkway to the second floor (see Fig. B). This change transferred all of the second

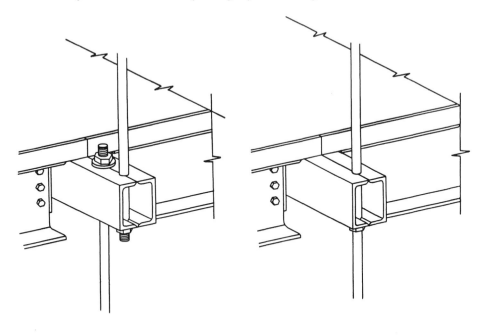

As Built   Original Detail

Figure B
Comparison of interrupted and continuous hanger rod details.
Source: *Investigation of the Kansas City Hyatt Regency Walkway Collapse*, NBS, U.S. Department of Commerce, p. 251.

floor load to the fourth floor box beam, doubling the load transmitted through the fourth floor box beam to the upper hanger rod. This submittal was stamped by the architect, structural engineer, and contractor, indicating their review.

The collapse occurred when the washer and nut on the upper hanger rod pulled through the fourth floor box beam, sending both platforms to the lobby floor with the fourth floor platform landing on top of the second floor platform. Even though a government investigation found that the original design was inadequate, it was felt that if the change had not been made the collapse would not have occurred. The judge held the structural engineering consultants liable for the accident, even though the engineers argued that the steel fabricators should be held responsible.

The judge based his ruling on the fact that the engineers as "licensed professionals" are responsible for assuring the structural safety of a building's design. He also stated that an engineer should not be allowed to "abdicate" his responsibility to another party, such as the steel fabricator. Further, the purpose of the shop drawing review process is to provide the opportunity for the engineering firm to verify the structural integrity of the design details.

This building failure illustrates the importance of good communication between the project participants, since any engineer or architect who took the time to review the impact of this change could have seen the possibility of a structural problem. Unfortunately, it appears that each reviewer stamped the submittal but assumed that someone else would complete the review.

Hyatt Walkway Collapse References

Investigation of the Kansas City Hyatt Regency Walkway Collapse. National Bureau of Standards, U.S. Department of Commerce, U.S. Government Printing Office, Washington, D.C., 1982.

"Hyatt Ruling Rocks Engineers." *Engineering News-Record*, McGraw-Hill, November 28, 1985.

"Hyatt Hearing Traces Design Chain." *Engineering News-Record*, McGraw-Hill, July 26, 1984.

With the construction phase nearing completion, the construction manager will get involved in monitoring and conducting all of the necessary tests needed to verify acceptance of the project. Elevators, compressors, mechanical and electrical systems, and so forth all need to be tested before they are officially accepted and payment is made. Before the owner can begin to use the facility, an official Certificate of Occupancy needs to be received. On major facilities the owner will have equipment and installation contractors whose work will need to be coordinated after the completion of the initial construction. It is not unusual for these "follow on" installation contractors to begin their work at the same time that the construction manager is overseeing the wrap-up of the last punchlist items.

The construction phase takes the project from procurement through the final completion of the facility and is a time when the bulk of the owner's funds will be spent. The amount of effort needed during this phase is dependent on how well the project team was prepared. If the scope/program of the project was well developed, the budget and schedule were realistic, the design documents correct, and competent, prequalified trade contractors selected, the construction phase should go smoothly. This preparation will keep change orders to a minimum, allowing the project team to focus on monitoring submittals and field work. As the final inspections are made and the operation of the project begins, the project moves on to the final project phase.

Project Closeout

The final step in the life of a project for the team is one of transitioning from design and construction to the actual use of the facility. This is likely to occur as the project begins to be utilized by the owner organization. As was mentioned above, the receiving organization may have their own contractors and facilities people who will be hard at work readying the facility for use as the project team closes out the project.

A major responsibility for the project team is to provide documentation for the end users. Shop drawings, warranties, guarantees, operation manuals, and as-built drawings need to be provided to the owner for use in the utilization and maintenance of the project. **As-built drawings** are original contract drawings adjusted to reflect all the changes that occurred; they define the project as it is being received. These drawings as well as the other documentation are information sources for the user organization's proper use and maintenance of the project. It is normal on large facilities for the user organization to assign people to the construction phase of the project for assistance in gathering all of this material as the project is being built. The installation contractors of major equipment may also be contractually required to provide training to user personnel on the operation and maintenance of the equipment.

The last step in the life of a project is the final accounting and assessment of the project team's performance. It is important for future projects that the team not repeat mistakes they may have made, just as their successes should be shared and carried on to the next project. Actual project and activity costs and durations should be recorded and compared to the project budget and should be used to update the historical records of the company. These updated costs will serve as the basis for the estimating and scheduling of future projects.

Conclusion

This chapter has followed a project from concept to completion. A construction management delivery method was assumed to show the interaction that can occur between owner, designer, and construction manager. The chapter has also attempted to identify and explain much of the terminology used in the industry. In the next chapter the delivery methods and contract types that are used will be explained, which will show how the project players can be brought together in different ways depending on the project type. Understand, however, that even though the players may be arranged contractually in different ways, the activities identified and explained in this chapter must still all be completed.

Chapter Review Questions

1. The identification and preparation of bid packages is an important coordination activity between the designer and construction manager.

   __ T   __ F

2. The term *fast-track* is synonymous with phased construction.

   __ T   __ F

3. In the construction phase of the project the owner needs to be most heavily involved.

   __ T   __ F

4. Value engineering is a process by which the quality of a project is reduced to save the owner money and time.

   __ T   __ F

5. A long-lead item is any material that takes considerable time from order to being received at the job site.

   __ T   __ F

6. The advertising for subcontractors, review of subcontractor's bids, and awarding of contracts occurs during which project phase?
   a. Procurement
   b. Construction
   c. Design
   d. Conceptual planning

7. The production of drawings and specifications is the output of which project phase?
   a. Procurement
   b. Construction
   c. Design
   d. Conceptual planning

8. A fair cost estimate is important for which of the following reasons?
   a. It identifies the fair cost of the work being contracted for.
   b. It serves as an opportunity for the owner, designer, and construction manager to familiarize themselves with the project.
   c. It can be used for projecting owner cash flow.
   d. All of the above.

9. Contractor prequalification involves evaluating a contractor with respect to which of the following?
   a. Financial condition
   b. Projects currently involved in
   c. Equipment owned
   d. Past project experience
   e. All of the above

10. As-built drawings, warranties, guarantees, and operation manuals are all provided to the owner during which project phase?
    a. Design
    b. Conceptual planning
    c. Construction
    d. Project closeout

Exercises

1. Using an actual local project, identify the major activities that occurred in each of the major project phases. Diagram these activities in the order that they occurred and identify which discipline accomplished each activity.

2. Using the same project as above, break the project down into appropriate bid packages. Write a scope of work for each bid package, identifying the major work to be accomplished in each.

CHAPTER

3

Construction Contracts and Delivery Methods

STUDENT LEARNING OBJECTIVES

From studying this chapter, you will learn:

   1. The risks inherent in a construction project
   2. The different delivery methods used to manage a project
   3. The different contract types used in the construction industry
   4. How to match delivery method and contract type to the project to best manage project risk
   5. What a contract change is and its impact on the project's contract

Introduction

The first chapter of this book described the construction industry, the types of projects, and the key players involved in a typical project. Chapter 2 presented a typical project and discussed how these players work together to manage such a project. It also identified the many tasks that must be accomplished, as well as when and by whom they must be performed. This chapter will look at how projects can be analyzed for risk, as well as the different ways that the project can be organized to minimize this risk. This chapter will discuss the different types of contracts, as well as the different methods that an owner can use to deliver a project. Contract changes will also be discussed.

Project Risk and Liability

An owner begins a typical construction project for the purpose of satisfying a particular objective. Chapter 1 has already explained how projects are typically unique undertakings, accomplished outdoors, and are large and consequently expose the owner to risk. What should be a clear, early goal of the project management team is to understand and assess what risks the owner is exposed to on the contemplated project. The risks that an owner faces are many, a few of which can be listed here:

The Project Site
Neighbors
Regulatory environment
Subsurface conditions
Economic climate

The Project
Project complexity
Planned technologies
Degree of finishes
Materials
Mechanical/electrical systems

The Process
Project funding
Timetable
Preconstruction information
Project unknowns

Owner Organization
Sophistication
Organizational structure
Decision making

The risks that an owner faces have been categorized into four groupings, the first of which considers the location where the project will be constructed. In analyzing the site it should be apparent that the more congested and politically sensitive the site and the more difficult the subsurface conditions of the site, the higher the risk to the owner. Inner-city sites, areas with economic uncertainty, and sites that have difficult subsurface conditions would all be characterized as having a high level of risk.

The type of project being undertaken also carries with it a certain degree of risk. In Chapter 1 the four different major categories of construction projects were described. One of the criteria used in establishing the project categories was the level of technology and to some degree the project's risk level. Projects that deal with known technologies, readily available materials, and an average level of finishes will carry less risk than those that require unknown technologies or unusual materials. Generally, industrial projects would be characterized as higher risks than residential projects.

The process that must be followed to complete a project can also carry with it a certain level of risk. To secure the necessary funding, for example, a developer may have to secure two major tenants by a specific date. Or, say, a new process facility must be completed by December 1 to maintain required levels of production. Any project that carries strict funding requirements, a rushed timetable, or has many unknowns will have a high level of risk. This is because commitments may have to be made with incomplete information or some of the early planning may have to be rushed.

The fourth category of project risk is the owner's organization. Sophisticated organizations with past project experience, a committed project team, and an empowered decision maker will be able to handle project risk better than unsophisticated owners, with little past project experience and no in-house construction expertise. On large institutional projects such as a hospital or a laboratory, where many organizational users will be involved, a strong, experienced decision maker is required to navigate the project through all the committees and users.

Chapter 2 described the large amount of fact-finding work required in the preconstruction stage of the project. Much of this work was needed to define the project's requirements (scope), as well as to understand the nature of the project for proper scheduling and estimating. It should be apparent that this research effort will also lead to a fairly accurate picture of the risk level for the project. This risk analysis should help to define the type of organization that will be necessary to manage the project, as well as give some indication of the amount of contingency needed to be built into the budget and schedule. **Contingency** is additional money or time added into a budget or schedule to allow for changes stemming from a better understanding of the project. The level of contingency is a direct measure of the degree of uncertainty on a project. Simply put, the greater the uncertainty, the more contingency is necessary.

In many ways risk is a factor of the unknown and the ability that the owner has to research, manage, and accept change. A congested building site and a highly complex and technical project coupled with a short timetable and an unsophisticated owner will always spell a high risk project. A project planned for a rural or open site, that is technically simple, and has good funding and a generous time frame to design and build will automatically provide less risk to the owner. Once the risks of the project are understood and identified, as well as the sophistication and capabilities of the owner, then the project team is ready to move to the next step—assembling the team necessary to deliver the project for the owner.

Delivery Methods

The term **delivery method** means the approach used to organize the project team so as to manage the entire designing and building process. The owner needs to decide which designers to hire, when to hire them, and under what type of contract. The owner also needs to decide when to hire the construction professional and under what type of contract. Which organization gets hired first? Do both organizations report to the owner, or does one report to the other? In some cases companies offer both design and construction services so that the owner only has to hire one company. There are a number of proven strategies that can be used to manage the process, each offering distinct advantages and disadvantages. This chapter will discuss the three most popular approaches: traditional, design-build, and construction management. Please note that other arrangements do exist, but these three are the most basic and, if understood, provide a foundation for comprehending the other more sophisticated delivery methods.

Traditional

In this arrangement the owner hires a design professional who prepares a complete set of contract documents for the owner for a design fee. With a complete set of contract documents in hand, the owner either negotiates a price with a general contractor or bids out the work. The general contractor is totally responsible for delivering the completed project as spelled out in the contract documents. The general contractor may subcontract out parts of the project, with each subcontractor reporting directly to the general contractor. The designer may be involved in overseeing the construction work in the field; this depends on the owner's needs and capabilities. In this delivery method no direct, formal relationship exists between the designer and the builder. They communicate only through the owner (see Fig. 3–1).

Advantages

A distinct advantage of this arrangement is that most owners, designers, and builders have worked under this framework on many projects and therefore are familiar with the system. Their control systems, documentation, and organizations are all set up to manage this process. The workers, subcontractors, and vendors also understand the system, improving overall job coordination.

Another advantage of this delivery method is that the owner can get a firm fixed price for the project before any work begins. Because an owner receives a complete set of contract documents before negotiating with or bidding out the work, it is reasonable to expect the contractor to provide a fixed price. As will be explained in the next section of this chapter, that price can be fixed and firm, or can be a cap with incentive clauses, or be based on unit prices. No matter which type of contract is used, the owner still has a very good idea as to the final price before the construction starts.

Related to that advantage is the opportunity to get good price competition from the open market. With a good set of contract documents the owner is able to advertise the job to everyone or, as was mentioned in Chapter 2, invite selected and prequalified bidders to price the project and then select the low bidder. Particularly in tough economic times, owners can get very good prices for their projects. A word of

Figure 3–1
Traditional approach.

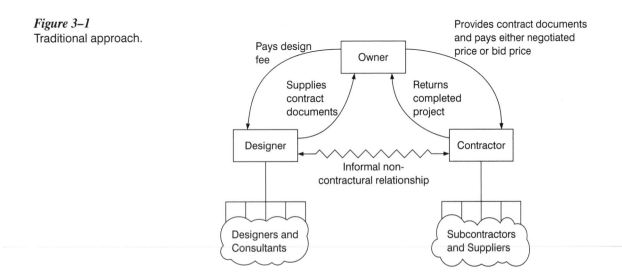

caution is that the contract documents must be accurate, containing no errors, and they must reflect exactly what the owner wants since any and all changes will be subject to renegotiation of the contract amount with the general contractor.

The final advantage is that in this arrangement the owner does not have to be actively involved on a day-to-day basis. The owner needs to be involved at specific review points during the course of the project, but does not need to play such an active role as in the construction management arrangement. Owner sophistication and a large design/construction staff are not necessary. The owner needs to be involved in selecting the designer and at formal review points, as well as in general contractor selection, but is not involved in the overall management of the process.

Disadvantages

One of the disadvantages of the traditional delivery method is that the contractors and subcontractors have no input until they are selected during the bid and award phase. It is possible for materials and methods to be specified that are not readily available or have been replaced by a more efficient process. Identifying long-lead items can be better done by the construction professional, and value engineering during design without contractor input could miss some key cost saving opportunities. Some design firms address this deficiency by hiring construction consultants, and the larger firms have construction professionals as part of their staff. Overall, the danger that must be guarded against is a design that is not constructable, or one that could be done better and cheaper, or a design that exceeds the owner's budget.

In the case of an owner wanting to competitively bid the project and receive a firm and fixed price, it is difficult to phase or fast-track the project. This arrangement, then, is the longest in terms of design and construction time. As was illustrated in the Fast-Track sidebar in Chapter 2, the design, bid and award, and construction all extend end to end with almost no overlap.

In this arrangement the owner, designer, and contractor all work autonomously. The designer designs the project based on owner instructions alone; the general contractor prices and schedules the project based on the construction documents alone. This approach provides little opportunity for interaction and team building between the participants, so that when interpretations have to be made, and they are made differently, major conflicts can occur. What accentuates this conflict is that the contractor-owner contract may be a firm, fixed price contract. Because the contractor may have competitively bid the job and in doing so had to interpret details as cost effectively as possible, his or her interpretation may very well be different from that of the owner and the designer. Differences in interpretation lead to conflicts which can quickly escalate, creating an adversarial relationship between all the parties.

Unforseen conditions can also be a source of conflict and lead to changes in the contract between parties. What is unforseen to one party (say, the contractor) may very well be assumed by another party (say, the designer). A thorough design process and a complete set of contract documents should attempt to minimize the number of interpretations, as described above, as well as attempt to eliminate as many unforseen conditions as possible. Conducting additional soil borings, opening up walls in renovation work, or calling in an asbestos consultant are the types of measures that should help to properly identify the actual conditions. Unfortunately, not every condition can be identified, and when unforseen conditions or events occur, the contract may very well have to be renegotiated.

Traditional Summary

The traditional arrangement as described has some distinct advantages as well as disadvantages. One of the major advantages is that the owner knows before construction begins what the cost of the project will be. To obtain this benefit, however, the owner gives up the ability to fast-track the project. The owner also gives up the benefits of contractor collaboration during the design phase. Another truth to realize is that the firm, fixed price is only as good as the contract documents. Changes to the scope of the project as well as design errors or omissions can lead to project delays and an increase in the contract price.

On many projects the time issue may not be important and the risks of changes in project scope may not be that great, so it may make sense to get good price competition and get a fixed price upfront. Projects that are not technically complicated or have been built before are candidates for this kind of arrangement. Many public projects are required to be built under this arrangement. As the complexity of the project increases due to the politics of the location, the technology of the project type, or the need to speed up the process, this arrangement begins to make less sense. A road paving operation, a single family home, or a warehouse could all be built using a traditional delivery method, but an emergency bridge repair or a commercial building project would have more success using another delivery method.

Design-Build

In the design-build arrangement the designer and the construction professional are either from the same company or through a joint venture, form a single company for the duration of the project (see Fig. 3–2). A **joint venture** is the legal binding of two companies for the purposes of providing a competitive advantage that would be difficult to attain alone. As an example, a design firm could form a joint venture with a construction company to offer a design-build service to an owner. In this arrangement the owner contracts with a single company early in the preconstruction stage, and this company takes the project from conceptual design right through construction. The terms **turnkey** and **design-manage** are also used to describe this process, although they actually are slightly different variations of the concept.

Advantages

One of the major reasons for choosing a design-build arrangement is to profit from the good communication that can occur between the design team and the construction team. Many of the largest design-build companies specialize in particular areas

Figure 3–2
Design-build approach.

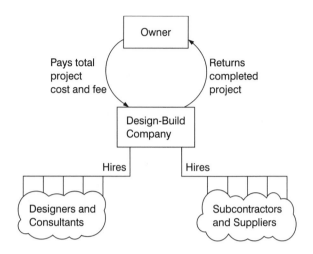

and have developed a smooth flow between the design and construction phases of the project. This collaboration allows the project to be easily fast-tracked, cutting down on the design-build time for the project.

Good communication between the designer and the construction professional also allows good construction input into the design phase, allowing for constructability analyses and value engineering. Good cost estimating and scheduling should occur throughout the entire project. Long-lead item identification and ordering should also be able to proceed smoothly.

In general, this arrangement allows easier incorporation of changes due to changed scope or unforseen conditions than in the other arrangements since the coordination is to occur within the same company. In this arrangement the owner is less heavily involved and is outside the communication between the designers and the builders. This keeps owner staffing to a minimum and speeds designer-builder communication.

Disadvantages

Although it is possible to give the owner a fixed, firm price before the project begins, such a procedure is generally not used in this arrangement. The reason is that the design-build firm is hired before the design for the project has been done, making the actual quantification and pricing of the project extremely difficult unless a past similar project has been done. Even if a firm price was given, the owner would still face the risk of the design-build firm sacrificing quality or scope to protect their profit. Generally the owner goes into this arrangement with a conceptual budget, but without a guarantee as to the final price. What makes this risk even greater is that if the project is fast-tracked the owner may not have a good idea as to the final price until part of the project is already built.

The fact that the owner has to be involved only minimally was listed above as an advantage, but it can also at times be a disadvantage. Because of the good communication and experience design-build companies have in working together, the project can move very fast. If the owner does not stay consistently involved throughout the process, he or she loses the ability to understand what is happening on the project, and when a decision needs to be made by ownership, all relevant factors may not be considered. Also, because of the speed with which the project is moving and the difficulty that the owner organization can have in keeping up with the project, it is easy for the work to proceed in a direction the owner may not want.

The last disadvantage regards the process of checks and balances. In the traditional arrangement the designer prepares a complete set of contract documents, which is used to measure and evaluate the performance of the contractor in the field. The designer is often hired by the owner to oversee the work in the field, and to ensure that deficient work is identified and corrected. In the design-build arrangement the designer works for the same company as the builder, so the design division of the company is therefore put in a position to critique/correct the work of the construction division, or, in a word, of itself. This is not quite as precarious as it seems since much of the work will be subcontracted to other companies and will be jointly managed by the design and construction professionals of the design-build company. The issue again is the lack of owner involvement and the dependency that the owner has on the quality and ethics of the design-build firm.

Design-Build Summary

The design-build process is one which benefits from the smooth coordination that can occur between the designers and builders within the same company. The owner gains the benefit of time and, taking the process to the extreme, can essentially tell the design-build firm what project is wanted, go on vacation, and return to a finished project. The negatives are total project cost is not generally known before construction begins, and the owner can easily get left out of the decision-making process, ending with a project that is not optimal for its use.

The design-build arrangement makes sense on projects of a highly technical nature where very good communication and coordination need to occur between the designers and builders. This delivery method also allows fast-tracking and is attractive to industries faced with strong competition and the need to get a new product to market quickly. The arrangement does not guarantee the best possible price, although subcontracts can still be competitively bid with some sharing of cost savings through an incentive clause, which will be explained later in this chapter. Projects that may use this arrangement include manufacturing plants, refineries, off-shore oil drilling platforms, and other technical projects that need good communication between the designers and builders and would benefit from a fast design-build process.

Construction Project Management

This delivery method was described in detail in Chapter 2. The owner hires both a design firm and a construction project management firm early in the preconstruction phase of a project. Which firm is hired first as well as which specific responsibilities

Figure 3–3
Construction project management approach.

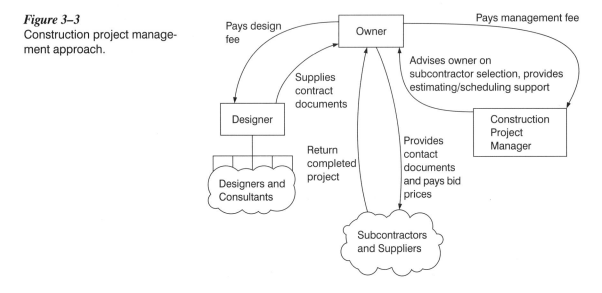

each firm will handle is variable and dependent on the level of involvement of the
owner as well as the expertise of the designer and construction professional (see Fig.
3–3). This delivery method has a number of variations such as Program Management,
Professional Management, Construction Management, and Professional Construction Management. The difference between these arrangements is a factor of the
expertise of the management team (i.e., whether the company is primarily a designer,
a builder, or a management consultant), when the management team is hired (i.e.,
concept, design, or procurement stage), and what responsibilities the owner assigns
to the different parties. As an example, the owner may want assistance through the
entire process, from programming right through project completion. This type of
assistance would include working with user groups and assisting in designer selection,
as well as overseeing the construction phase. This would typically be called a program
management delivery method. In a construction management delivery method the
owner would do much of the programming and designer selection alone, and look to
the construction manager to do the work as outlined in Chapter 2. The advantages
and disadvantages of these delivery methods are similar and are outlined below.

Advantages

One of the major advantages of this arrangement is that good communication is
established early in the designing/building process between the owner, designer, and
construction professional and continues through the completion of the project. The
process encourages collaboration, allowing the construction people to critique and
influence the design of the project, just as the designers have a part in contractor
selection and in reviewing the work in the field. This cooperation between the entire
designing/building team leads to a good value engineering program, which is one of
the greatest attributes of this delivery method.

This arrangement allows phasing since the design and construction people are able to get together early and develop the necessary coordination schedules.

Another advantage of this delivery method is that the owner receives the cost benefit of the competition between the subcontractor bids. In this arrangement the subcontractors are under contract to the owner, so that if the project is broken down into 20 bid packages with 5 bidders per package, the owner receives the benefit of 100 competitive bids. The designer and construction manager review and recommend the contractor, and the owner receives the financial benefit (see Fig. 3–4).

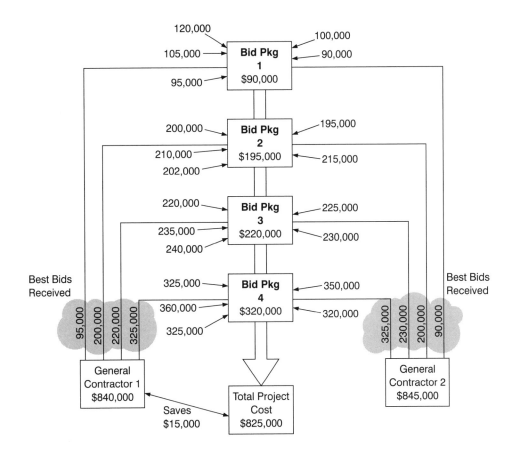

By soliciting more bids and by passing the savings directly to the owner, the owner saves $15,000 compared to the best general contractor price. This example illustrates four bid packages. Typical commercial building projects may have thirty or more bid packages.

Figure 3–4
Cost savings due to competitive bidding.

Furthermore, the implementation of changes during the course of construction is not as difficult as in the traditional method since the designer and construction manager are in close communication. Ideally, the team should be able to anticipate changes, minimizing their impact on the project.

Disadvantages

For this arrangement to work well good communication and cooperation need to exist between the owner, designer, and construction manager. If any of the players become inflexible, uncooperative, or uncommunicative, all of the advantages listed above can quickly become disadvantages. This delivery method is heavily dependent on the shared, mutual respect between the players, respect which will be repeatedly tested.

High owner involvement is necessary in the construction management arrangement for this delivery method to work. In principle the designer, construction manager, and owner form a team with each taking on different shared and individual responsibilities and become dependent on each other. This arrangement requires a more sophisticated owner than is required in the other two delivery methods. Please note that this is how project management evolved. Some owners desired a construction management approach but either did not have the sophistication or the time needed, and consequently hired a project manager to take their place. The project manager essentially becomes the owner and in some cases may even hire a construction management company, creating a second tier.

This delivery method has a tendency to encourage fast-tracking since the construction management team gets involved early in the process. On some projects phasing the job can be risky because the scope of the project may be variable or project financing still questionable. If the team ignores these risks and pushes the project into construction anyway, the owner can incur significant financial penalties.

Construction Project Management Summary

This arrangement offers significant advantages to the owner provided the owner is willing and able to stay active in the process and selects a good designer and construction project manager who are willing to work as team players. This delivery method offers the cost advantages of competitive bidding to the owner, as well as the opportunity to phase the project.

This delivery method is commonly used by real estate developers in the commercial building industry. In some cases, the project will start with a construction project management arrangement, and as the design nears completion the construction manager will negotiate a fixed or guaranteed price with the owner and the project will become a traditional arrangement. On very large "mega" projects, like the Alaskan pipeline, a program manager will be hired by the owner to manage the entire program. Their job is to take this very large project and break it down into smaller packages which will be designed and constructed by separate design and construction companies.

Student Union Example

Your university has received a large private donation targeted to build a new student center. The university has recently completed a master plan and has an ideal site selected for locating the new building. Many on campus have strong opinions as to what the facility should look like (particularly the Architecture Department), as well as what services should be located in the building. How does the university organize to manage this opportunity?

The major risks involved in the student union project are the schedule, the organizational structure, and the project environment. The project is not technically difficult, and a good amount of information exists on similar projects to adequately plan this project. The site soil conditions are well documented and should not pose a problem. The schedule is critical because the university would like to complete and dedicate this new student union at the 75th anniversary of the school. This creates a very tight total timeframe for the project! The organizational structure of any university is a problem because of the many sub-units that exist. The faculty, students, and administration all have their own organizations and should be involved in the process. To do this requires a team of people who understand the university and can involve the different groups at the correct times. The university exists in a fairly dense urban environment with the building site in proximity to a residential neighborhood. This requires that the city and the neighborhood be involved.

A construction project management arrangement was chosen as the delivery method for this project for a number of reasons:

1. This method allows the project to be phased, given the tight design-build period to accomplish the project.
2. The university, with its physical plant staff, assigned project architect, consulting architect, and construction manager should be able to work with the university organizations to involve the right people at the right times.
3. The same project team, by being involved early, should also be able to properly investigate the environment of the project and properly work with the neighborhood and the city.

Any phased project carries the risk of the project beginning construction without a completely defined project cost, but given the need to complete the project early and the fact that this project is not technically complicated, the university was willing to absorb that risk.

Contract Types

In addition to choosing a delivery method for a project, the owner must also decide what type of contract to use. A **contract** is simply an agreement between two or more people in which one person agrees to perform a specific task or provide goods or a service to another in exchange for something in return. The contract type chosen, like the delivery method chosen, is important to the owner for its ability to address project risk. In this chapter three basic types of contracts will be discussed: single fixed price, unit price, and cost plus a fee.

Single Fixed Price

In a single fixed price contract, also called a **lump sum**, the contractor agrees to provide a specified amount of work for a specific sum. In this contracting method both parties try to fix the conditions of the project as precisely as possible. Once the contract is signed, both parties must live with its terms.

The advantage of this contracting method is that the owner knows before the work begins what the final cost of the project will be. This contracting method is usually used in the traditional delivery method described earlier in the chapter. The designer will prepare a complete set of contract documents, which the owner then either bids out or negotiates with a contractor. A final contract amount is agreed to and the work begins.

The risk that the owner takes in this contracting method is that the contract is only as good as the accuracy of the contract documents—if the scope of the project changes or if errors exist in the documentation, the contract will need to be renegotiated, possibly exposing the owner to increased financial risk.

As was also explained in the traditional delivery method, to allow a fixed price contract to be negotiated, a complete set of contract documents must be prepared. This takes time and prevents the construction of the project from beginning until the design work is complete. This negates the possibility of a fast-tracked project.

In summary, this contracting method combined with a traditional delivery method allows the owner to define and commit to an agreed-upon project description and dollar amount before the work begins. For owners who want to minimize risk on a project that can be clearly defined (i.e., that has minimal unforseen conditions), this type of contracting method works well. The owner must understand that the process will take longer and that changes caused by mistakes, unknowns, or changes in owner requirements will jeopardize the agreement.

Unit Price Contract

In a unit price contract the owner and the contractor agree as to the price that will be charged per unit for the major elements of the project. The owner/designer will typically provide estimated quantities for the project, then ask contractors to "bid" the job by calculating unit prices for these items and calculating a final price. Contractor overhead, profit, and other project expenses must be included within the unit prices that are provided. The owner then compares the final prices and selects the low bidder (see Fig. 3–5).

The advantage of this type of contracting method is that in many projects (heavy engineering projects being a perfect example), it is difficult to accurately quantify the work necessary. In excavation work it is often difficult to accurately quantify the actual amount of rock versus earth that must be excavated. To eliminate risk to both the owner and the bidders, the designers will estimate quantities and then ask the bidders to provide a unit price for each type of excavation and bid the job. Actual payments will be made on the basis of multiplying the actual quantities excavated by the unit price provided.

This contracting method provides the owner with a competitive bid situation, allowing for a fair price for the work. It also eliminates the risk of getting a fixed price and then having to renegotiate because of differing site conditions, as explained above. In this contracting method, work can also begin before the design is completed, speeding up the completion of the project.

The risk to the owner in this contracting method is if estimated quantities are significantly different from the reality of the situation, the financial commitment of the owner may be greater than planned. Mistaken estimates also expose the owner to what is called an **unbalanced bid**, increasing the project's costs to the owner (see Fig. 3–6). Of note is the fact that significantly unbalanced bids border on being considered unethical today, and in some cases can be rejected or the unbalanced work items can be deleted by a change order.

In this contracting method actual quantities must be measured in the field, requiring an owner presence on site to work with the contractors. Delivery tickets

Work Items	Unit	Estimated Quantity	Bidder 1 Unit Price	Bidder 1 Bid Amount	Bidder 2 Unit Price	Bidder 2 Bid Amount
Soil Excavation	CY	10,000	5.50	55,000	2.00	20,000
Rock Excavation	CY	3,000	25.00	75,000	25.00	75,000
6″ Pipe	LF	600	17.00	10,200	18.00	10,800
Crushed Stone Fill	CY	4,000	21.00	84,000	20.00	80,000
Fill Material	CY	6,000	14.00	84,000	20.00	120,000
Top Soil 4″ Deep	SY	400	5.00	2,000	6.00	2,400
TOTAL				$310,200		$308,200

Bidder 2 wins the job with the $308,200 total price.

Figure 3–5
Unit price example.

	Estimated Quantity	Bid Price	Actual Quantity	Amount Paid
Soil Excavation	10,000	20,000	8,000	16,000
Rock Excavation	3,000	75,000	3,000	75,000
6" Pipe	600	10,800	600	10,800
Crushed Stone Fill	4,000	80,000	4,000	80,000
Fill Material	6,000	120,000	7,000	140,000
Top Soil 4" Deep	400	2,400	400	2,400
				Total $324,200

Assume Bidder 2, in Figure 3.5, knew that the soil excavation quantity provided was high and the fill material quantity provided was low. By providing a low unit price for soil excavation and a high unit price for fill material, Bidder 2 earns an additional $16,000.

Figure 3–6
Unbalanced bid (Bidder 2).

and other invoices must be checked and validated. Final contract price is not known until the last item of work is measured and invoiced by the contractor.

In summary, heavy engineering projects such as earth dams, dredging operations, and underground utility work are often accomplished by a unit price contract since the quality of the work can be defined, but the actual quantities are difficult to determine in advance. The risk that the owner runs on this type of contract is that the actual price is not known until the work nears completion, but can be minimized by good design support. For example, good subsurface exploratory work can help predict actual quantities in advance. An owner presence in the field must also exist to verify quantities and authorize payments. Once a good estimate is made of the actual quantities and funding is deemed adequate, work can begin before final design is complete, saving project time.

Cost Plus a Fee

In a cost-plus contract arrangement, also called a **reimbursable** or a **time and materials** contract, the contractor (or, sometimes, the designer) works on the project and is reimbursed by the owner for its costs, plus is paid either an additional agreed-upon fee or is paid a fee which is a percentage of those costs. It is important for the owner to spell out clearly in advance what costs will be reimbursed and which costs are to be covered by the fee.

This contract makes sense when the scope of the project may be difficult to define or when it is important to fast-track the project. By using this type of contract the contractor can start work without a clearly defined project scope since all costs will be reimbursed and a profit guaranteed. This type of contract also allows the contractor, designer, and owner to work together early in the designing/building process in a nonadversarial fashion, encouraging value engineering and good estimating and scheduling support.

A variation of this type of contract is called a **guaranteed maximum price (GMP)**. In this type of contract the contractor is reimbursed at cost with an agreed-upon fee up to the GMP, which is essentially a cap; beyond this point the contractor is responsible for covering any additional costs within the original project scope (see Fig. 3–7). It

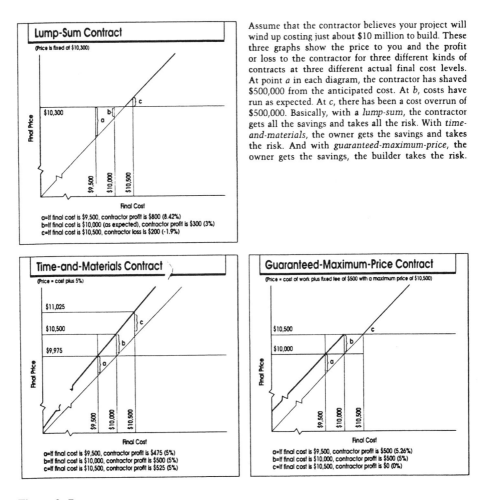

Assume that the contractor believes your project will wind up costing just about $10 million to build. These three graphs show the price to you and the profit or loss to the contractor for three different kinds of contracts at three different actual final cost levels. At point *a* in each diagram, the contractor has shaved $500,000 from the anticipated cost. At *b*, costs have run as expected. At *c*, there has been a cost overrun of $500,000. Basically, with a *lump-sum*, the contractor gets all the savings and takes all the risk. With *time-and-materials*, the owner gets the savings and takes the risk. And with *guaranteed-maximum-price*, the owner gets the savings, the builder takes the risk.

Lump-Sum Contract
(Price is fixed at $10,300)

a=If final cost is $9,500, contractor profit is $800 (8.42%)
b=If final cost is $10,000 (as expected), contractor profit is $300 (3%)
c=If final cost is $10,500, contractor loss is $200 (-1.9%)

Time-and-Materials Contract
(Price = cost plus 5%)

a=If final cost is $9,500, contractor profit is $475 (5%)
b=If final cost is $10,000, contractor profit is $500 (5%)
c=If final cost is $10,500, contractor profit is $525 (5%)

Guaranteed-Maximum-Price Contract
(Price = cost of work plus fixed fee of $500 with a maximum price of $10,500)

a=If final cost is $9,500, contractor profit is $500 (5.26%)
b=If final cost is $10,000, contractor profit is $500 (5%)
c=If final cost is $10,500, contractor profit is $0 (0%)

Figure 3–7
Cost vs. price contract comparison.
Compliments of John D. Macomber, the George B. H. Macomber Co., Boston, MA

is not unusual to include in this contract an **incentive clause** which specifies that the contractor will receive additional profit for bringing the project in under the GMP.

The risk to the owner in using this type of contract is that, even with a GMP, the project is started with considerable unknowns. By using a GMP the project costs may be capped, but the quality and scope may become sacrificed at the expense of the GMP. If a GMP is not used, the scope and quality of the project may be solid, but the cost and schedule may increase. This type of contract requires a reputable contractor or construction manager, since tremendous trust needs to be placed with this participant.

In summary, the cost-plus type of contract makes sense when the owner needs to complete a project quickly or when the project is difficult to define accurately upfront. The project needs a qualified and reputable designer and builder, as well as an active owner organization. The risks to the owner are clear: because the work often begins before the project is completely defined, the costs may very well exceed the figures that were defined upfront. A GMP can provide a cap, but this cap may be protected by the contractor at the expense of quality and scope. This type of contract is used in both the construction project management and design-build delivery methods.

Contract Changes

The previous section described the primary types of contracts utilized in the construction industry. In each situation a contract is established between the owner and the contractor stating that a certain service or material will be provided for a stipulated fee. This contract is entered into after the work to be performed has been estimated and scheduled and a work plan has been established. Depending on the contract type, a bid may be submitted or negotiations may occur between the owner and the contractor. The bid or negotiations are based upon the design to date and the best knowledge available at the time about the project. Then sometimes the situation changes.

Contract changes occur for three main reasons:

1. Because of a change in owner requirements, the scope of the project changes.
2. Because of conditions unforseen at the time the contract is signed, the work must be performed differently.
3. Due to omissions or design features that cannot be built as specified, the design must be adjusted.

The impact that the change has on the contract depends on which type of contract is in place and what the reason is for the change. A cost-plus contract can accommodate all of the above contingencies without a change in the contract except for possibly the case of a GMP. If a GMP is in place, it may have to be increased, depending on the terms of the GMP clause between owner and contractor. In the case of a fixed price contract, all three of the above reasons will probably lead to a change in the contract between parties. All of the above generally lead to increased costs and time which need to be fairly adjusted. In the case of a unit price contract, the reasons listed above may or may not lead to a contract change. In the case of an excavation project an increase in rock would be covered by the unit price submitted, whereas an unexpected decision to prohibit blasting would require a change in the contract.

Contract changes are a reality on construction projects, although from the perspective of most parties they can be disruptive and should be avoided. As illustrated above the type of contract chosen can either increase or decrease the number of changes that need to be negotiated. In general, fixed price contracts require the most, and cost plus a fee the least. Owners need to recognize that changes cost the project money since in negotiating a change with a contractor they will generally not get as good a price as if they had included the change item in the original project, where the work may have been competitively bid.

Student Union Example Continued

Both the architect and the construction manager were hired using cost plus a fee contracts. This type of contract was best for both of these professionals since it was necessary to involve both players early in the designing/building process before the project, and their roles, could be clearly defined.

The trade contractors who bid on the work were hired using single fixed price contracts. The bid packages were 100 percent complete and were competitively bid on by four to six prequalified contractors per bid package. A single fixed price contract was possible since the architect and construction manager were able to put together complete construction documents for each of the bid packages.

Conclusion

The purpose of this section (Chapters 1–3) is to provide a broad overview of the construction industry and the construction project. This section is designed to set the stage for the remainder of the book, which will focus on the tools that the project manager needs to use to manage a project. At this point the student should have a good understanding of the construction industry, its players, the major activities that need to occur throughout the life of a project, and the different ways that the project players can come together. The student should understand what is meant by project risk and how the different delivery methods and contracts can be used to minimize project risk for an owner.

The next section of this book will discuss estimating—why estimates are performed and how they are performed. It should be clear at this point that good upfront estimates are critical to the success of a project.

Chapter Review Questions

1. The delivery method that an owner chooses should be in response to the amount and type of risk that an owner sees in a project.

    ___ T   ___ F

2. The greater the risk that an owner sees in a project, the lower the contingency that needs to be applied.

    ___ T   ___ F

3. A delivery method is a type of contracting method.

    ___ T   ___ F

4. A joint venture is the legal binding of two companies for the purpose of providing a competitive advantage that would be difficult to provide alone.

    ___ T   ___ F

5. Contract changes are more likely to occur on a single fixed price contract than on a cost plus a fee contract.

    ___ T   ___ F

6. The advantage(s) of a traditional delivery method is (are):
    a. Reduced project time
    b. Nonadversarial relationships between participants
    c. Known project cost before construction
    d. All of the above

7. The developer of a 40 story high-rise office building desires the shortest possible construction time. What delivery method would be best?
    a. Traditional
    b. Design-build
    c. Construction project management
    d. All of the above
    e. B and C only

8. Which of the below listed reasons is *not* a cause of a contract change?
    a. A change in owner requirements
    b. Unforseen conditions
    c. Designer omissions or errors
    d. Poor job site productivity

9. Which of the below would be a source of owner project risk?
   a. Project complexity
   b. Environmental regulations
   c. A short designing/building timeframe
   d. A complicated owner organization
   e. All of the above

10. Which type of contractual arrangement would be best used when the quantities of work are difficult to determine in advance?
    a. Single fixed price
    b. Unit price
    c. Cost plus a fee
    d. None of the above

Exercises

1. Identify the different delivery method options for the following situations:
   a. A supermarket chain wants to build a new store.
   b. A hospital needs to build a new wing to add new diagnostic equipment. Ongoing operations must be continued.
   c. A family wants to build a single vacation home.

   In all situations identify the risks associated with each project and the delivery options. Recommend a solution.

2. Suggest a contract type for each of the three project types in Exercise 1. Then, by combining the delivery method and the contract type, identify the risks that the owner has assumed and avoided for each project.

Sources of Additional Information

Barrie, Donald S., and Boyd C. Paulson. *Professional Construction Management*, 3d ed. New York: McGraw-Hill Inc., 1992.

Clough, Richard H., and Glenn A. Sears. *Construction Project Management* 3d ed. New York: John Wiley & Sons, Inc., 1991.

Hendrickson, Chris, and Tung Au. *Project Management For Construction*. Englewood Cliffs, NJ: Prentice Hall, 1989.

Hinze, Jimmie. *Construction Contracts*. New York: McGraw-Hill Inc., 1993.

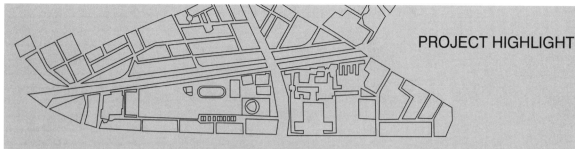

MIT Renovation of Building 16 and 56

SECTION ONE

Defining the Project and the Team

Nancy E. Joyce, Senior Program Manager, Beacon Construction

Buildings 16 and 56 sit side by side in the center of Massachusetts Institute of Technology's campus along a corridor spine that connects most buildings on their lower levels. The two buildings are simple rectangular structures of eight floors with penthouse space above and mechanical space in the basement and subbasement areas. Building 16 was built in 1952 and Building 56 in 1965. Combined, they have an area of 250,000 square feet. They originally housed Food Nutrition Science, but over the years evolved into more mixed use, housing Chemical Engineering, Chemistry, and Biology classrooms and administrative areas. Neither building has had a major renovation in its lifetime.

With the completion of a new Biology building, two-thirds of the occupants of Buildings 16 and 56 moved over to the new building. This presented MIT with the opportunity to renovate these buildings and provided space to relocate occupants of Building 20, a three-story "temporary" wood timber structure, of World War II vintage. Building 20 sits in the middle of what will be a new development site for MIT (see Fig. A). Located in the middle of an urban environment, there is little opportunity for expansion beyond the boundaries of the campus. Removing Building 20 will provide the Institute with the opportunity for better use of existing land within the campus.

Before a project team was assembled, MIT explored the appropriate level of renovation needed to accomplish the Institute's goals. These goals were to demolish Building 20, to consolidate the departments left in Buildings 16 and 56, and to create a facility that would meet today's standards for laboratories. In a Facility Assessment Study, conducted by a consultant, the conditions of the building systems were examined, and renovation alternatives were developed and priced. The study

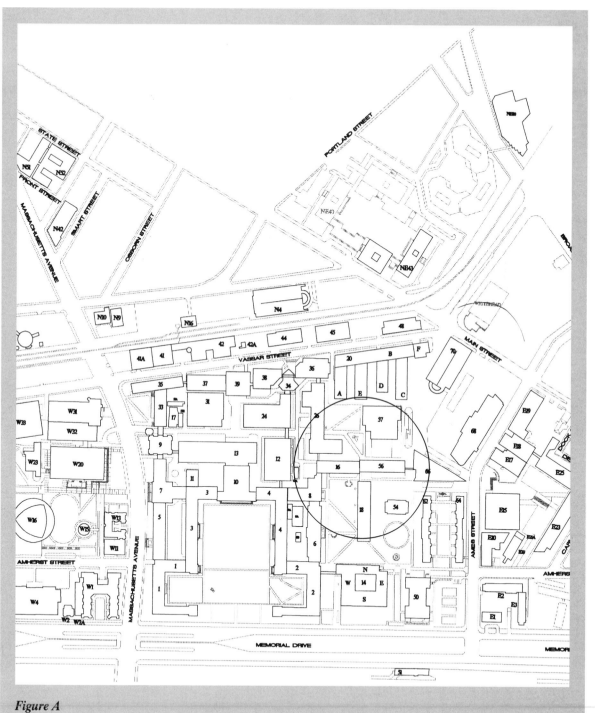

Figure A
Partial map of MIT, showing the location of Buildings 16 and 56.

explored three levels of work: demolition of all or part of the buildings with new construction on the site, major renovation and replacement of existing systems, and repair of existing systems with cosmetic upgrade. Because the cost of building new was considerably higher than renovating and because the logistics of a major demolition and new construction in the center of campus would be disruptive, the alternative of new construction was not considered viable. Repair of existing systems, it was shown, would only buy a short amount of time before major work would have to be done and it would not provide the opportunity to bring the facility up to today's laboratory standards. In the end, the owner decided to renovate the buildings and replace the existing systems. This solution would provide the Institute with new systems and a facility set up for another 25 to 30 years of functional life.

For MIT the project was unique in many ways. The scope of the renovation was ambitious and the proposed uses diverse. User groups to be relocated to the buildings reported organizationally up different chains of command. This complicated unilateral decision making and created the potential for customization of spaces during design. Users who occupied the buildings had to be maintained during construction, which created technical complications. Also, identifying and coordinating the reuse of building systems added another dimension to the design and construction process.

Given all the unique aspects of the project, MIT gave careful consideration to how to formulate a project team. Because this was a renovation with technical and programmatic complexities, the Institute decided to build the project utilizing the services of a construction manager. They also hired the construction manager for preconstruction services. This ensured that constructability issues were addressed during design and that cost and schedule were integrated in all design decisions. When hiring both the construction manager and the architect, the owner looked for experience in laboratory buildings. Laboratory design is a specialized field and requires extensive understanding of mechanical systems. Since mechanical systems can make up 60 percent of construction costs, the engineering of these systems is a critical aspect of any design. MIT was also concerned about the integration of their own campus standards and operating procedures into the design. Because these two buildings sat in the middle of an existing group of research buildings, ensuring the continuity of these standards and procedures was critical. The architectural firm and the construction management firm hired by the owner both had extensive experience with laboratory design and construction and had worked on projects at MIT over the years and were familiar with academic decision processes. The two firms had also established a good working relationship on a recently completed project at MIT. In addition, the construction manager also had prior owner experience and was able to provide support in the handling of some of the more traditional owner functions such as user group interface, furniture inventory, hazardous waste management, permit applications, and moving coordination.

At the time the team was established, the final occupancy was not yet determined. Because new codes dictated a much higher movement of air through the laboratories than they were originally designed for, it was unclear what the buildings would support in terms of new laboratory use. If the buildings were to be used heavily as lab space, it became clear that new penthouses would have to be built to house the extra air handlers needed. After examination by the structural engineer, it was found that the additional penthouses would necessitate structural bracing throughout the buildings to accommodate the additional loads the equipment would add to the building. Because of the extensive cost associated with this, the decision was made to include only as many laboratories as the building could support without the addition of any penthouse space. There were cost advantages to following this avenue, but the tradeoff was limited future growth for the laboratory users in the buildings.

The decision to limit laboratory usage helped to determine the final occupancy. Once this was determined, the architects were able to begin the design phase of the project. One of the Institute's goals was to empty Building 20, which necessitated an ambitious program. Users were programmed, with their existing square footage being the maximum amount of space they could occupy in the new building. This was done by the Institute with the belief that efficiencies in the new buildings would offset any growth needs. In addition, the buildings would be set up with common support spaces on each floor, thereby reducing the number of individual support spaces needed. The aggregate of these decisions was expected to alleviate any program crowding. During the schematic design stage these assumptions were tested and, with few exceptions, proved correct. At the conclusion of schematic design the broad scope of the project was defined. With approvals from the various user groups, and an approved project budget and schedule and a final review by the owner, the architects were authorized to move to the more detailed phase of design development. Meetings with the user groups were very focused during this phase. Attention was given to the details of how each lab operated. The project conference room was outfitted with mock-ups of different building systems, graphics of similar finished spaces, and real measurements mapped out on the floor. These props were useful for people not accustomed to visualizing space. At the conclusion of these meetings and with approvals of the users, the project team again reviewed the project with the Institute. With a confirmed project budget and schedule, the architect moved into the construction documents stage. This was the production phase for the architects. The scope and design decisions had been worked out during design development, and, barring any significant changes, the users were not actively involved in the construction documents phase.

During the time the architects were designing the various phases of the project, the construction management team was pro-actively monitoring the cost and schedule impacts of various design decisions, exploring construction alternatives, conducting constructability analysis, and advising the owner on technical issues. When the architects were producing construction documents, the construction team was isolating Building 56 from the rest of the campus and separating systems that serviced other buildings. The team was also preparing Building 56 for demolition by cleaning debris out of the building , identifying and salvaging reusable equipment, furniture, and casework, cleaning out the asbestos and other hazardous materials accumulated through 30 years of laboratory use, building temporary pathways around the site for MIT personnel, and setting up a construction site.

At construction start, the project team—consisting of the owner, the architect, and the construction manager—was firmly established and all members were familiar with the Institute's policies and concerns on the project. The designer and the construction manager had prior experience with the owner, both collectively and separately. Most issues that inevitably arise when forming a new team were resolved and the roles of each member were understood as we accelerated into construction.

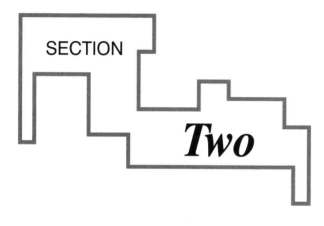

Estimating

This section of the book addresses the subject of estimating. It looks at the reasons why estimates are an important project management tool and explores the different types of estimates that are used. You will see that as the project evolves, the estimates evolve too. More is being learned about the project, its scope is being clarified, decisions are being made, consultants are being brought in to offer expert advice, and schedules are being produced. All of this activity feeds information to the estimators and allows them to produce more accuracy in their estimates.

In Chapter 4 the fundamentals of estimating are covered. The chapter looks at the information needed to develop the different types of estimates, the length of time needed to develop them, and the percentage of accuracy that each type of estimate will provide. Chapter 5 covers estimates that are used during the conceptual and design phases. These are called rough order of magnitude estimates, square foot estimates, and assemblies or systems estimates. Chapter 6 covers detailed estimates, which are used during the bidding and award phase.

CHAPTER

4

Estimating Fundamentals

CHAPTER OUTLINE

From studying this chapter you will learn:

1. The reasons why estimates are used
2. The types of estimates that are used
3. The basic issues that affect project prices

Introduction

The question often asked when discussing the subject of estimating is, "Is estimating an art or a science?" I would answer that it involves a little of both. A good artist has the ability to visualize and anticipate, is creative, and can provide answers to questions never before asked. A scientist is methodical, organized, and technically strong, has strong research abilities, and can perform complex calculations. A good estimator must possess strong organizational and communication skills, particularly as the project increases in size and complexity. Preparing an estimate is expensive, involving many people throughout the organization. On a large project (over $200 million) a company might spend as much as $500,000 on estimating. A good estimator needs the ability to visualize the project, to think multi-dimensionally. As an example, consider a construction activity like painting. Painting while standing on a floor can be done fast, up high on a ladder or scaffold at a slower rate, or on a ladder while working around the flooring contractor at an even slower rate. An estimator needs a firm grasp of every detail to estimate the painting costs accurately.

Estimators work both from experience and from anticipation. They are logical in that they study past similar projects and research how long an activity lasted and how much it cost, and then they consider new variables: What are the current technologies? What is the time of year? How aggressive is the schedule? Out of this process comes the new estimate. An estimator must consider the many variables that affect estimates: quantities (how much), productivity (how long), weather and strikes (unforseen conditions), and overhead and profit (factors of market conditions).

A good data base of past project experiences is essential to preparing a quick and accurate estimate. Professional estimators spend considerable time and resources developing and protecting this data base. Each new project provides a clearer picture of the actual cost of construction and adds to the value of the data. Successful design and construction companies own these data bases, as do independent cost consultants and cost data suppliers such as R.S. Means, which sell construction cost information to owners, designers, and constructors.

Why Estimate?

Just as the name implies, an estimate is an educated guess, an appraisal, an opinion, or an approximation as to the cost of a project prior to its actual construction. This estimate can be prepared at many points during the life of a project, as will be explained shortly. Still you can ask, why is an estimate important anyway?

From an owner's perspective, an early estimate serves to answer important questions such as the following:

1. Is the project affordable?
2. How large a project can be constructed for the money available?
3. What level of quality can be included in a project?
4. Which project options make the most sense?

Identifying costs early on facilitates sound decision-making but since that estimate will be prepared early in the project's life, it will be prepared with little "hard" design information.

Estimates also provide guidelines to the designer. As a project is being designed, it is important that the designer select materials and size the project within the budget of the owner. As the project proceeds through the design phase, the design must be continually compared to the owner's budget. If the design of a project begins to exceed the budget, the designer must determine the best alternatives for cost reduction. Estimating and designing are highly related, as a change in either forces a change in the other. As computers become more sophisticated as a project management tool, designers will be able to specify a material item such as a valve and be able to instantaneously review the impact of this decision on the project cost.

At the end of the design process estimates must also be prepared by the individual trade contractors to figure their bid price. These estimates are done with design documents complete or nearly complete and are the most time consuming and most accurate of the estimates. It is also a good idea for the owner/designer/construction manager team to prepare a detailed estimate at this point to verify the accuracy of the bid prices and to negotiate with the trade contractors. This is the time in the life of a project when the owner clearly begins to know the actual cost of a project.

Types of Estimates

When beginning to prepare an estimate, it is important to understand its intended use. An estimate can be prepared at any point throughout the life of a project. Depending on the information available and the time spent preparing the estimate, the accuracy it provides will vary (see Fig. 4–1).

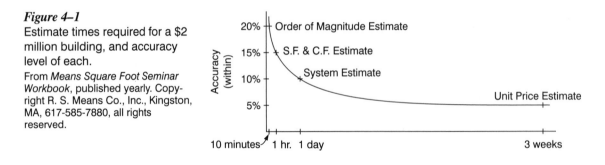

Figure 4–1
Estimate times required for a $2 million building, and accuracy level of each.
From *Means Square Foot Seminar Workbook*, published yearly. Copyright R. S. Means Co., Inc., Kingston, MA, 617-585-7880, all rights reserved.

Conceptual Phase

An owner needs cost information very early on in a project so that decisions as to the location and scope can be made before money is spent on design or property purchase. This estimate will be prepared with very little information, relying mostly on historic data and whatever descriptions are available. This type of estimate is called a conceptual or rough order of magnitude and is generally prepared with a construction start several years away.

The description of a project may be a sketch or a brief written description. The size of the project is generally known, although it may be described in terms of capacity such as the number of beds for a hospital, of pupils for a school, or megawatts for a power plant. The time needed to prepare this type of estimate is short, generally in the range of a half day or less, and the presentation is generally informal for the purpose of providing a target budget. Estimates are often prepared for many different program options so that the best alternative(s) can be selected.

Schematic Phase

As the project moves into the schematic stage the designer and possibly the construction manager (depending on the delivery method chosen) have become involved in the design and estimating of the project. The program for the project has been provided by the owner, and the project team may be incorporating different design alternatives into the basic design. A schematic estimate will generally be based on a design that is approximately 30 percent designed and includes the following information:

- 1/16″ floor plans, elevations, and sections
- Outline specifications for most trade sections
- One-line drawings for mechanical and electrical systems

The preparation of this estimate includes some area takeoff, and calculating of the major project elements such as the gross floor area of a building, the exterior wall area, or the gross cubic yards of earth to be excavated. At this stage some of the key subcontractors might be asked for input with respect to the more complicated systems. This estimate involves the use of some unit pricing combined with the use of assemblies and gross square foot costs. On a major commercial building project this estimate will take one to two weeks and will carry a 10 percent contingency. A contingency is added to allow for the unknown design and engineering details that will be developed during the next design stage, as well as the evolution of the project scope.

At the end of the schematic design stage the presentation of the design to the owner is accompanied by an estimate of the cost of the project. Any design alternatives will also be accompanied by estimates. Before the project team moves on to the next phase of design, the owner will decide on the basic design parameters and on the project budget. Any cost reduction ideas will be presented and priced by the estimators. Some of these ideas may be accepted or rejected at this stage, and some may be carried forward to be better defined in the next phase. The estimator also will identify the major assumptions that form the basis of the estimate. These assumptions are understood as the project scope at the schematic design stage and will be used as a reference point as the project moves into the next phase.

Design Development Phase

The estimate that is put together at the design development stage is prepared similarly to the schematic phase estimate. The difference is that the level of information is much more defined. Because of this, the time to prepare the estimate is longer but the accuracy is greater. A design development estimate will generally be based on a design that is 60 percent complete and includes the following information:

- Drawings shown at ⅛″
- Elevations, sections, and details at a larger scale
- All relevant specifications sections
- Mechanical and electrical systems well defined

Most of the major project items will be quantifiable, and the more important unit prices should be known at this point. Depending on the delivery method chosen, either key trade subcontractors or key consultants will be involved in the pricing of the more complicated systems. By this stage a network schedule will have been begun, allowing a better understanding of the overall duration of the project as well as when each of the major project elements is to be constructed. The preparation of this estimate should take two to three weeks and be accurate to within 5 to 10 percent of the final cost.

With the presentation of this estimate the costs of the materials and methods will be known and should be compared to past similar projects. If any are significantly higher or lower than normal, they should be examined. Major assumptions should be noted and compared to what was assumed at the schematic design stage. Because the design is further advanced, fewer assumptions will be needed at this point. It is a good idea to compare this estimate by cost category with the schematic estimate, and to investigate any significant variances. Remember that the estimate at this stage is a tool. It should be used to verify a design within the owner's budget and to identify any good cost savings ideas.

Procurement Phase

For the purposes of this discussion a traditional delivery method will be assumed using a lump sum bid. With this arrangement, an estimate would be prepared by all the contractors who are bidding the work, as well as by the project team. The contractors prepare the estimate to identify a price to bid, and the owner team prepares an estimate to be in a position to negotiate a fair price and to verify the accuracy of the contractors' prices. This estimate is prepared based on a complete set of contract documents.

The contractors bidding the project will break the job down into work packages (see the sidebar in Chapter 2) and request bids from prequalified subcontractors for each package. Most general contractors will do some of the work with their own workforces and therefore will not request bids in these areas. Estimates done for bidding require a complete understanding of material quantities, which is taken from the drawings, and thoroughly researched unit prices, which usually involves input from local material suppliers. An exact schedule will be prepared. This will be used to identify the duration of the project, knowledge of which is needed to accurately estimate general conditions items (see sidebar on Direct vs. Indirect Costs). Depending on the size of the project, a bid estimate can take three weeks or longer to prepare. These detailed estimates are extremely accurate; the difference between bidders is often only the profit margin that they are including.

The estimate prepared by the owner team, sometimes called a **fair cost estimate,** will actually be prepared by the construction manager or a professional cost estimator. It relies more on in-house expertise for its accuracy and less on subcontractors and suppliers. This procedure will provide less accuracy than the bid estimate, but enough accuracy to serve as a check on the bid price. Normally, the estimate done at the design development stage will be updated based upon design decisions. However, good quantity take-offs and accurate unit pricing will be incorporated, and a schedule will also be prepared.

The bids that are received will be on a standardized bid form and will be reviewed for accuracy and completeness by the project team. The advantage of having prepared a fair cost estimate is that the owner team is already knowledgeable about the details of the design and in a good position to scrutinize the contractors' bids. The completed bids should be summarized by divisions and should identify sales tax, project overhead, home office overhead, and project profit.

Sidebar ▬▬▬▬▬▬▬▬▬▬▬▬▬▬▬▬▬▬▬▬

Direct vs. Indirect Costs

The costs involved in the construction of a project can be broken down into two major categories—direct and indirect. Direct costs are the costs associated with the purchase of building materials and the labor associated with the physical installation of these materials. The cost of roofing material, the purchase of asphalt, or the cost of landscape material would all be considered a direct cost. Also, the rental of a paving machine with its operator, the daily wage of a carpenter, and the costs associated with the finishing of concrete would all be considered direct costs. Direct costs essentially occur in the field, and once work is stopped on a project, the incurring of direct costs stop.

Direct costs are not the only costs that accrue on a project, since for a company to stay in business and bid and win a project it must incur additional costs. These overall corporate expenses are generally incurred at the home office and are called home office overhead. Examples of these costs would be executive salaries, legal and accounting fees, office rental, vehicle expenses, and clerical fees. These costs are incurred even if there are no projects under construction at the time.

There are also indirect costs that occur in the field. These costs are called general conditions or field office overhead and are the costs that are necessary to supervise and support the job site. Examples of these costs would be the rental of the job site trailer, the superintendent's salary, and the cost of security fencing, a guard, or signage.

Direct costs are calculated by researching the unit costs of the materials and labor being utilized and determining the quantities required. The larger the project, the greater its complexity, and the higher the quality, the greater the direct costs.

Indirect costs are more a factor of the project's duration and the degree of supervision required. As a project's duration is increased or if a project requires a high level of coordination, the indirect costs of a project correspondingly increase. As will be covered later in the control section of this book, project managers need to determine a project duration that minimizes the combined costs of the project. See the discussion of optimum project duration in Chapter 10.

Estimate Considerations

Every estimate, whether it is generated in the conceptual phase of a project or at bid time, must consider the same basic issues. Project price is affected by the size of the project, the quality of the work, the location, the construction start and duration, and by other general market conditions. The accuracy of an estimate is directly affected by the ability of the estimator to properly analyze these basic issues.

Project Size

As a general rule in construction, as a project gets bigger the cost of the project increases. The size of the project is a factor of the owner's needs. Size is handled differently depending on the stage in the project's life at which the estimate is being conducted. At the conceptual stage, size will be more an issue of basic capacity, such as apartment units for a real estate developer or miles of roadway for a highway engineer. As the project becomes a little better understood, the project's size will begin to be quantified more accurately. The basic capacities will begin to be thought of in terms of more specific parameters such as square footage of floor or roof, numbers of on and off ramps, or quantity of excavation. Further design leads to more specific quantities, eventually ending with exact numbers for every project item. When quantities are difficult to determine, a unit price contract is used, as was discussed in Chapter 3, and the bidders will then provide unit prices, and the actual quantities will be measured during construction.

The principle of **economy of scale** must be considered when addressing project size. Essentially, as projects get bigger they get more expensive, but at a less rapid rate. This occurs because the larger the project, the more efficiently people and equipment can be utilized. Also, as people repeat a task, particularly many times over, they get better and faster at it, reducing the cost of labor. On large commercial building and heavy engineering projects worker productivities are plotted into what are called **learning curves** (see Fig. 4–2). It shows that as the number of repetitions or units a worker needs to accomplish increases, the time required to perform that repetition decreases. Estimators treat project size by establishing tables which recognize the typical size of a project and a respective price and then adjusting up or down accordingly from this norm. An example of how this is done is provided in Chapter 5.

Project Quality

As the quality and complexity of a project increases, so does the project's cost. A high level of quality may be required for aesthetic reasons as specified by the project archi-

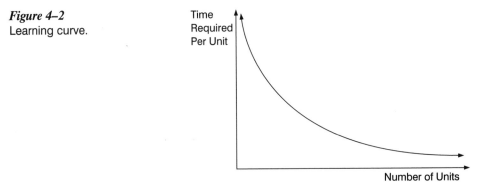

Figure 4–2
Learning curve.

tect, or the quality may be required for the safety of the project users or the public. A nuclear power plant, a Titan launch facility, or a corporate board room all might require a higher level of quality than normal. When ascertaining quality, the estimator must consult governmental regulations, the end users, and the project designers.

In considering project quality, it must be recognized that as the expected quality of a project increases, the cost of providing this quality increases as well, but at a progressively greater rate. Related to this is the fact that as the quality of a project increases, the user experiences increased project satisfaction, but at a lesser rate (see Fig. 4–3). This chart identifies the importance of arriving at the optimum level of quality for the project, since to specify an increased level of quality beyond what is required can increase the project's cost substantially while not providing a corresponding value to the client.

An example of this might be in the specification of a crane system for an assembly plant. The owner needs a 200-ton lifting capacity and the ability to maneuver the load to within ½ inch. The plant foreman has noted that the floor operations would be speeded up if the precision of movement of the crane was halved to ¼ inch. The value to the owner of providing more precise movement must be compared to the cost of providing it. In this case the ¼-inch precision may only be provided by one crane manufacturer, a circumstance which increases the cost tremendously. The cost of the specification must be estimated and compared to the increased productivity benefit.

The tools that the estimator uses to estimate quality get more refined as the design of the project becomes better understood. In the early stages of a project the

Figure 4–3
Project cost vs. value.

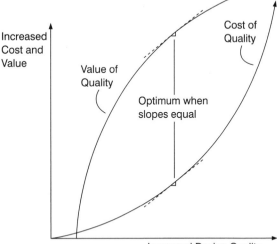

The optimum level of design quality is the point at which the slope of the two curves is equal. Beyond that point the cost of providing one more unit of value exceeds its corresponding value.

Figure 4-4
Project quality by quality levels.

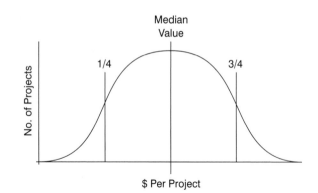

estimator must compare the project to other past projects broken down into quality levels such as ¼–median–¾ (see Fig. 4–4). In this example, a ¾ project cost would signify a project quality level with 75 percent of projects of a lesser quality and 25 percent more expensive. A ¼ project cost would indicate that 75 percent of similar projects would be of higher quality. As the project becomes further designed, the designer will begin to specify materials and systems each with corresponding material and installation prices. The estimator at this stage must now quantify the work required and the corresponding prices. As the estimator moves to the bidding stage of a project, quality must be precisely quantified per individual unit. This is one reason why a detailed estimate takes longer to prepare and requires more clearly defined contract documents.

Location

Where the project will be constructed is a major consideration in the preparation of an estimate. Depending on the location, a great variation exists in the purchasing of materials and their delivery, rental or purchase of equipment, and in the cost of labor. Material costs are a factor of availability, competition, and access to efficient methods of transportation. Labor costs, particularly unionized labor, is a factor of the strength of the local bargaining unit. The cost of labor is also a factor of the degree of sophistication and level of training found at the project location. On some projects (the Alaskan Pipeline was a good example) the number and the skill levels of workers required are not available locally, so labor forces have to be imported.

The cost of constructing projects in different locations can be predicted by establishing what are called **location indices** for different cities and parts of the country. An index is created for a particular city by comparing the cost of labor, equipment, and material for that city to the national average. This allows an estimator using national average costs to adjust the estimate to a particular location. An example of how to adjust for different locations is presented in Chapter 5. Most major design and construction companies have developed an accurate set of location indices which they use for their pricing, or they buy this cost data from national pricing suppliers.

Time

When a project is built, just like where it is built, can have a major impact on the cost of the project. Since estimates, by definition, are prepared in advance of the physical construction, the estimator must "project" to the future what the cost of the work will be. Moreover, the estimate must predict what the cost of material and labor will be *when these costs will be paid*—not when the estimate is prepared. Initial project estimates are often prepared two or more years in advance of the start of construction, and if the project takes three years to construct, the estimator therefore must identify costs as far as five years into the future.

Publications such as the R. S. Means data books and *Engineering News-Record* do a good job of tracking actual project costs by the use of **historical indices** (see Fig. 4–5), a similar concept to the location index just discussed. These indices allow a project's historical cost to be adjusted to today. An example of how to adjust a project using historical indices will be worked through in Chapter 5. This adjustment

Table 13.2-011 Historical Cost Indexes

Year	Quarterly City Cost Index Jan. 1, 1993 = 100 Est.	Quarterly City Cost Index Jan. 1, 1993 = 100 Actual	Current Index Based on Jan. 1, 1995 = 100 Est.	Current Index Based on Jan. 1, 1995 = 100 Actual	Year	Quarterly City Cost Index Jan. 1, 1993 = 100 Actual	Current Index Based on Jan. 1, 1995 = 100 Est.	Current Index Based on Jan. 1, 1995 = 100 Actual	Year	Quarterly City Cost Index Jan. 1, 1993 = 100 Actual	Current Index Based on Jan. 1, 1995 = 100 Est.	Current Index Based on Jan. 1, 1995 = 100 Actual
Oct 1995					July 1980	62.9	59.5		July 1962	20.2	19.1	
July 1995					1979	57.8	54.7		1961	19.8	18.7	
April 1995					1978	53.5	50.6		1960	19.7	18.6	
Jan 1995	105.7		100.0	100.0	1977	49.5	46.8		1959	19.3	18.3	
July 1994		104.4	98.8		1976	46.9	44.4		1958	18.8	17.8	
1993		101.7	96.2		1975	44.8	42.4		1957	18.4	17.4	
1992		99.4	94.1		1974	41.4	39.2		1956	17.6	16.7	
1991		96.8	91.6		1973	37.7	35.7		1955	16.6	15.7	
1990		94.3	89.2		1972	34.8	32.9		1954	16.0	15.1	
1989		92.1	87.2		1971	32.1	30.4		1953	15.8	14.9	
1988		89.9	85.0		1970	28.7	27.2		1952	15.4	14.6	
1987		87.7	83.0		1969	26.9	25.4		1951	15.0	14.2	
1986		84.2	79.7		1968	24.9	23.6		1950	13.7	13.0	
1985		82.6	78.2		1967	23.5	22.2		1949	13.3	12.6	
1984		82.0	77.5		1966	22.7	21.5		1948	13.3	12.6	
1983		80.2	75.8		1965	21.7	20.5		1947	12.1	11.4	
1982		76.1	72.0		1964	21.2	20.1		1946	10.1	9.6	
▼ 1981		70.0	66.2		▼ 1963	20.7	19.6		▼ 1945	8.8	8.3	

To find the **current cost** from a project built previously in either the same city or a different city, the following formula is used:

$$\text{Present Cost (City X)} = \frac{\text{Present Index (City X)}}{\text{Former Index (City Y)}} \times \text{Former Cost (City Y)}$$

For example: Find the construction cost of a building to be built in San Francisco, CA, as of January 1, 1995 when the identical

building cost $500,000 in Boston on July 1, 1968.

To Project Future Construction Costs: Using the results of the last five years average percentage increase as a basis, an average increase of 2.5% could be used.

The historical index figures above are compiled from the Means Construction Index Service.

$$\text{Jan. 1, 1995 (San Francisco)} = \frac{\text{(San Francisco) } 105.7 \times 127.0}{\text{(Boston) } 24.9 \times 120.7} = \$500,000 = \$2,233,500$$

Figure 4–5

Historical indices can be used to adjust the cost of a past project to one today.

combined with the location adjustments covered above allow an estimator to estimate the cost of a new project today in one location by looking at a similar project built several years ago and hundreds of miles away (Fig. 4–6).

Unfortunately, it is difficult to project with accuracy what the index will be for a future year, so the best an estimator can do is look at current trends and anticipate future labor and material prices.

Phoenix   Boston

Figure 4–6
By the use of location indices and historical indices, the cost of a new skyscraper in Boston can be projected by adjusting for location and time from a similar structure built five years ago in Phoenix.
Photos by Don Farrell

Other Market Conditions

An estimator who accurately incorporates the four major considerations just identified—project size, project quality, location, and time—will have an estimate that
reflects the fair value of the project. Assuming a normal market without any unusual
circumstances, this estimate should reflect the price to be paid.

Market conditions, however, shift, and owners, designers, and contractors all look
at a given project from a different perspective. In a market without much work, contractors may bid a project at cost or with little profit to cover their overhead and keep their
staff employed. On complicated projects contractors may bid the work low in the hopes
of making additional profit on future change orders or new work. It is not unusual for
contractors to provide very competitive prices when they look to enter a new market or
establish a relationship with a new owner. Some owners and designers are viewed as difficult to work with and may not receive good prices, particularly if the market is strong.

The above issues are difficult to quantify, but should be considered in the preparation of the estimate. These factors are usually treated as a percentage applied at
the end of the estimate, included in either overhead and profit or in a final contingency. Good investigatory work in the project area as well as discussions with interested contractors can help an estimator better understand these factors (see Successful Construction Estimating sidebar).

Conclusion

This chapter has discussed the reasons why estimates are prepared and how and by
whom they are used. It should be clear that an estimate is not static but evolves with
the project and should be used as a tool to provide the owner a better product. Estimates are first used in the conceptual stage of a project and, because of little available information, are only able then to provide a rough idea as to the project's cost,
but as the design evolves and more and more is understood about the project, the
estimate becomes progressively more accurate. During design the estimate can be
used to evaluate systems and products and is an integral part of the value engineering process. After the design is completed the estimator is able to provide a complete and accurate estimate that reflects the true value of the project. This detailed
estimate uses unit prices and when used by the contractor is called a bid estimate.

Also covered in this chapter were the basic factors that every estimate must consider: project size, project quality, location, time, and other market issues. Project
size must be understood to determine quantities needed, whereas project quality
must be specified to obtain unit prices. Location can be treated by either researching local prices or by the use of specific area cost indices. Project time and market
considerations are more difficult to quantify since both must look to the future and
are dependent on economic as well as local issues. Projecting past trends to the
future is often the best way that these two factors can be forecast.

Sidebar ▬▬▬▬▬▬▬▬▬▬▬▬▬▬▬▬▬▬▬▬▬▬▬▬▬▬

Successful Construction Estimating

A construction estimate is a forecast of a project's "actual" cost. I am oftentimes asked in a classroom setting, "What is the correct answer?", referring to a class estimating exercise. "Correct answers" in construction estimating are only known after a project is complete and "actual" costs are totaled. Success in construction estimating is correctly forecasting a project's "actual" cost.

The next best approach to actually building a project to arrive at the "actual" cost is to *visualize* building the project through the estimating process. Building the project in your mind's eye, or visualizing the process, is fundamental to achieving *realistic* estimate totals. To do otherwise will most often result in overlooked construction steps or components that, all totaled, could be significant.

The estimating process consists of breaking a project down into logical components (e.g., excavate for spread footings, form up for spread footings, place concrete for spread footings), which are then **scoped, quantified,** and **priced**. The summation of the components along with other overhead and profit issues would then comprise the project estimate.

Scoping: The scoping of project components is one of the first steps in the estimating process. Each component is made up of unique dimensions, specified quality and construction methodology, and potential problems and solutions. Through scoping and visualizing the scoping, an understanding of the component makeup is achieved. As an example, the scoping of a project's strip footing consists of understanding that it is continuous around the perimeter of the building, 3000 PSI concrete is required, it will be placed 2 feet below finish grade, that soils data indicates rock will be encountered in the northwest corner, work will be done during the winter months, 3 each continuous #4 reinforcing bars are required, and the perimeter will be difficult to reach with direct chute from the concrete truck. Without the above scope, items important to quantifying and pricing might be missed, such as the need for a pumper truck for the concrete placement, pricing out 3,000 PSI concrete, concrete winter protection, and the potential need for doweling into rock.

A project's scope is derived from design documents (plans and specifications), and the estimator's experience and construction background. Project site visits are encouraged for addressing such scoping issues as site access/egress, storage capabilities, utilities locations, and the extent of ongoing operations. This is particularly important in scoping repair/remodeling projects.

Quantifying: Quantifying is the packaging of project components' scope into units that can be priced. In the example above, quantities would consist of "cubic yards" of concrete, "pounds" of reinforcing steel, "square feet" of strip footing formwork, and "days" rental of a concrete pumper truck.

Pricing: Pricing involves applying marketplace labor, material, and equipment costing to the quantities. Care should be taken to apply pricing that is relevant to the project location, quality, and job specifics such as the wage rate requirements on government projects.

Overhead and profit issues: Other project estimate issues that must be addressed over and above the project component scoping, quantifying, and pricing is the contractor's home office overhead, profit, sales taxes, labor burden, bond, and contingency. A construction estimate is not complete without a review of these "adders," which are generally applied as a function of the total bare construction direct costs.

Rory Woolsey, President
The Wool-Zee Co., Inc.
Bellingham, Washington

Chapter Review Questions

1. As estimates progress from conceptual to detailed, they take longer to prepare and get less accurate.

    ___ T   ___ F

2. Conceptual estimates work with fewer but larger units than a detailed estimate.

    ___ T   ___ F

3. On a $2 million building project a systems estimate should take about a day.

    ___ T   ___ F

4. Estimating and value engineering are unrelated processes.

    ___ T   ___ F

5. To prepare a conceptual estimate, completed drawings and specifications are required.

    ___ T   ___ F

6. An estimate would *not* be used for which of the following reasons?
    a. To determine how large a project can be funded
    b. To determine what level of quality can be funded
    c. To determine the project's duration
    d. To evaluate different project options

7. A contractor would prepare which type of estimate to bid a project?
   a. Conceptual
   b. Square foot
   c. Assemblies
   d. Unit price

8. The concept of a learning curve is as follows:
   a. Repetition and productivity are unrelated.
   b. As repetitions increase, productivity increases.
   c. As repetitions increase, productivity decreases.
   d. Worker productivity is a constant.

9. As the design quality of a project increases, the value of the quality to the owner _____.
   a. increases at an increasing rate.
   b. increases at a decreasing rate.
   c. increases at a constant rate.
   d. decreases.

10. Which of the following would *not* have an impact on the cost of a project?
   a. Market conditions
   b. Contractor exposure to a new market
   c. Contractor exposure to a new client
   d. All of the above would have an impact on project cost

Exercises

1. In the PBS series video *Skyscraper* (see your school video library) the projected cost of the project changed over time. What were some of the reasons why the price changed? Was the early, conceptual estimate wrong? Could the cost of the project have been better controlled? What should have been done differently?

2. Using *Engineering News-Record* (ENR) as a source, plot the Construction Cost Index (every three months is adequate) over time from January 1985 to January 1995. The Construction Cost Index is found on the Market Trends page. Assuming a project cost $1 million in January 1985, how much would the same project cost in:
   a. June 1986
   b. September 1988
   c. June 1994

   Use your graph to calculate your answer. List five reasons why the actual cost might vary from your estimate.

Conceptual, Square Foot, and Assemblies Estimating

From studying this chapter you will learn:

1. How to prepare a conceptual estimate
2. How to prepare a square foot estimate
3. How to organize, develop, and utilize an assemblies estimate

Introduction

This chapter will focus on the types of estimates that are prepared during the preconstruction stage of a project. These estimates would be prepared by the owner, designer, and/or construction manager depending on the delivery method that has been chosen. Chapter 4 provided an overview of these estimates; this chapter will go through the steps involved in the preparation of each of them.

As was discussed in the previous chapter, there are several factors to consider when preparing an estimate: the size of the project, the quality of the materials and methods used, the project location, the time of the year, and the market conditions. As the project proceeds, more information becomes available about all these factors. As a consequence, the time required to prepare the estimate increases, but the accuracy of the estimate also increases.

All preconstruction estimates start with a data base of past projects. Companies that are large or have been in existence for several years normally develop a history of project costs, which is used for the estimate of future projects. Companies that are new, too small, or looking to enter a new market may have to purchase this data from companies that specialize in the research and sale of project costs. The data presented in this chapter comes from the R.S. Means company's Assemblies Costs and Square Foot Costs books. Both of these books are published annually; they reflect the actual cost of construction in the publication year on a national basis and are available to the public for purchase.

In working with historical data it is important to understand what costs are included and not included in each line item. Are costs separated by project type? Is overhead and profit included or broken out? What type of foundations were used? What was the quality level of the project? What was its location? All of these questions and more must be considered as the estimate is prepared.

Conceptual Estimate

Approach

Conceptual—also called **rough order of magnitude** (ROM)—estimates are typically developed by establishing a cost per usable unit from past projects and multiplying this cost times the number of units being proposed. An example of these costs might be cost per bed for a hospital, cost per apartment, cost per pupil for a school, or cost per mile for a highway. If the costs are developed on a national average basis, they must be adjusted using the appropriate city cost index. Costs taken from past projects must also be adjusted to current or future dollars. If the proposed project will be smaller or larger than normal, the cost can also be adjusted for size. Lastly, an appropriate contingency should be applied to allow for scope adjustments as well as economic or market conditions. Conceptual estimates can be done quickly, in 10-15 minutes, and provide an accuracy in the ±20 percent range.

Data

The accuracy of a conceptual estimate is dependent on the quality of the data that the estimator has available. The best scenario would be to look into the company data and find the exact project size, quality, and location, then adjust for inflation and market conditions, and the estimate is done. Unfortunately, unless you happen to be estimating stores for a major chain, most projects vary enough that it is difficult to compare one with another. Companies that specialize in certain areas of work often do have reasonably good data on that type of project, but companies that do a lot of different kinds of work must rely on published data (see Fig. 5–1).

The data illustrated in that index is based on an average of over 11,500 projects as reported to Means from contractors, designers, and owners. These costs are all adjusted to the current year and averaged. The unit costs are divided into three columns, ¼, median, and ¾. This allows the estimator to consider and adjust for quality quickly. The median cost represents the cost at which 50 percent of the projects surveyed are more expensive and 50 percent are less expensive. The ¾ value would represent a higher quality project, where only 25 percent of comparable projects were more expensive.

After selecting the appropriate project type and quality value, the next step is to multiply the cost times the appropriate number of units. Let's take the example of a 50-unit motel; assume a high quality (3/4) price, 1995 data:

number of units × unit cost = total cost
50 units × 40,700 per unit = $2,035,000

14.1 000 | S.F. & C.F. Costs

			UNIT	UNIT COSTS			% OF TOTAL			
				1/4	MEDIAN	3/4	1/4	MEDIAN	3/4	
500	0100	Site work	S.F.	5.35	7.80	13.05	8%	11.60%	16.10%	500
	1800	Equipment		1.20	1.97	3.23	2.10%	2.90%	4.60%	
	2720	Plumbing		3.04	4.27	5.40	6.80%	9%	11.50%	
	2730	Heating, ventilating, air conditioning		1.62	3.15	3.36	4.20%	6%	6.40%	
	2900	Electrical		2.70	3.88	5.55	5%	6.50%	8.10%	
	3100	Total: Mechanical & Electrical		8.15	11.75	16.15	15.60%	19.20%	23.50%	
	9000	Per apartment, total cost	Apt.	47,600	53,000	67,100				
	9500	Total: Mechanical & Electrical	•	7,800	10,700	13,400				
510	0010	ICE SKATING RINKS	S.F.	40.30	70.40	97.50				510
	0020	Total project costs	C.F.	2.81	2.88	3.32				
	2720	Plumbing	S.F.	1.43	2.27	2.68	3.10%	3.20%	5.60%	
	2900	Electrical		4.10	4.52	6.30	6.70%	10.10%	15%	
	3100	Total: Mechanical & Electrical	↓	7.25	10.45	13.05	9.90%	25.90%	29.80%	
520	0010	JAILS	S.F.	115	144	192				520
	0020	Total project costs	C.F.	10.70	12.70	17.20				
	1800	Equipment	S.F.	4.87	12.95	22.50	4%	8.90%	15.10%	
	2720	Plumbing		11.45	15.10	18.35	7%	8.90%	13.30%	
	2770	Heating, ventilating, air conditioning		10.50	14.30	26.50	7.40%	9.40%	17.70%	
	2900	Electrical		12.80	16.60	20.40	7.90%	11.30%	12.40%	
	3100	Total: Mechanical & Electrical	↓	30	43.30	57.15	24.30%	30.10%	31.70%	
530	0010	LIBRARIES	S.F.	69.55	87.70	109				530
	0020	Total project costs	C.F.	4.96	6	7.55				
	0500	Masonry	S.F.	3.91	7.15	13.75	4.50%	7.40%	10.70%	
	1800	Equipment		1	2.60	4.31	1.10%	2.80%	4.80%	
	2720	Plumbing		2.80	3.93	5.35	3.40%	4.60%	5.60%	
	2770	Heating, ventilating, air conditioning		6.20	10.20	13.25	8.70%	11%	14.60%	
	2900	Electrical		7.15	9.30	11.90	8.20%	10.90%	12.10%	
	3100	Total: Mechanical & Electrical	↓	14.85	19.55	28.75	17.10%	22%	28%	
550	0010	MEDICAL CLINICS	S.F.	68.10	84.15	106				550
	0020	Total project costs	C.F.	5.20	6.70	8.85				
	1800	Equipment	S.F.	1.82	3.88	5.85	1.80%	4.80%	6.80%	
	2720	Plumbing		4.71	6.50	8.70	6.10%	8.40%	10%	
	2770	Heating, ventilating, air conditioning		5.60	7.20	10.60	6.60%	9.20%	11.30%	
	2900	Electrical		5.95	8.30	11	8.10%	10%	11.90%	
	3100	Total: Mechanical & Electrical	↓	14.40	18.35	25.80	18.60%	23.60%	29.60%	
	3500									
570	0010	MEDICAL OFFICES	S.F.	63.35	79.60	98.20				570
	0020	Total project costs	C.F.	4.87	6.65	8.95				
	1800	Equipment	S.F.	2.15	4.26	6.05	3%	5.80%	7.10%	
	2720	Plumbing		3.69	5.65	7.65	5.60%	6.80%	8.60%	
	2770	Heating, ventilating, air conditioning		4.35	6.50	8.30	6.10%	8%	9.50%	
	2900	Electrical		5.15	7.55	10.50	7.60%	9.70%	11.60%	
	3100	Total: Mechanical & Electrical	↓	11.95	16.70	22.35	16.60%	21.20%	26%	
590	0010	MOTELS	S.F.	41.80	60.60	76.15				590
	0020	Total project costs	C.F.	3.59	5.06	8.30				
	2720	Plumbing	S.F.	4.17	5.20	6.35	9.40%	10.60%	12.50%	
	2770	Heating, ventilating, air conditioning		2.20	3.79	5.55	4.90%	5.60%	8.20%	
	2900	Electrical		3.88	4.90	6.35	7.10%	8.20%	10.40%	
	3100	Total: Mechanical & Electrical	↓	9.40	12.05	16.60	18.50%	23.10%	26.20%	
	5000									
	9000	Per rental unit, total cost	Unit	21,000	31,500	40,700				
	9500	Total: Mechanical & Electrical		4,025	5,700	6,150				

SQUARE FOOT

Figure 5–1

Sample square foot costs for various structures.

From *Means Assemblies Cost Data 1995*. Copyright R. S. Means Co., Inc., Kingston, MA, 617-585-7880, all rights reserved.

100

This total reflects the cost of building a typical, high quality motel in a national average location. It includes contractor's overhead and profit and, assuming this project is of a normal size, should represent the average of the bids received in that year, under normal market conditions.

Adjustments

Staying with the motel example, there are adjustments that may have to be made. The first adjustment is for size. As a project increases or decreases from what the data is showing as average, the cost per unit increases or decreases to reflect economies in buyout and worker efficiency. In general, larger projects can be built more efficiently than smaller projects because materials can be bought in larger lot sizes and worker productivity generally increases as workers "learn" the job. Referring to Figure 5–2, the typical high quality (3/4) motel unit would be 620 square feet. Therefore:

motel's size = 620 sq.ft. × 50 units = 31,000 sq.ft.

S.F. & C.F. Costs	**R14.1-030**	**Space Planning**

The figures in the table below indicate typical ranges in square feet as a function of the "occupant" unit. This table is best used in the preliminary design stages to help determine the probable size requirement for the total project. See R141-050 for the typical total size ranges for various types of buildings.

Table 14.1-031 Unit Gross Area Requirements

Building Type	Unit	Gross Area in S.F.		
		1/4	Median	3/4
Apartments	Unit	660	860	1,100
Auditorium & Play Theaters	Seat	18	25	38
Bowling Alleys	Lane		940	
Churches & Synagogues	Seat	20	28	39
Dormitories	Bed	200	230	275
Fraternity & Sorority Houses	Bed	220	315	370
Garages, Parking	Car	325	355	385
Hospitals	Bed	685	850	1,075
Hotels	Rental Unit	475	600	710
Housing for the elderly	Unit	515	635	755
Housing, Public	Unit	700	875	1,030
Ice Skating Rinks	Total	27,000	30,000	36,000
Motels	Rental Unit	360	465	620
Nursing Homes	Bed	290	350	450
Restaurants	Seat	23	29	39
Schools, Elementary	Pupil	65	77	90
Junior High & Middle		85	110	129
Senior High		102	130	145
Vocational		110	135	195
Shooting Ranges	Point		450	
Theaters & Movies	Seat		15	

Figure 5–2
Unit gross area requirements.

In the Project Size Modifier table (see Fig. 5–3), the typical motel equals 27,000 sq.ft. Therefore, the cost multiplier for the size adjustment can be calculated as follows:

$$\frac{\text{Proposed building area}}{\text{Typical building area}} = \text{Size factor}$$

$$\frac{31,000}{27,000} = 1.148$$

Using a size factor of 1.148 (round off to 1.1), read from the graph in Figure 5–3 a cost multiplier of .99.

Therefore for the motel project the price, adjusted for size, equals:

Base cost × Cost multiplier = Size adjusted cost
$2,035,000 × .99 = $2,014,650

This price reflects the cost of a high quality motel, adjusted to a slightly lower unit price since the motel is slightly larger than normal. Another adjustment that may have to be made is for location. The price that is shown above, $2,014,650, reflects the cost of building this motel in a national average city with a construction start in early 1995. Adjustments can be made for location by comparing the price of common building materials and labor from one city to another.

Figure 5–4 shows tables for a few of the over 200 cities throughout the United States and Canada for which Means has compiled the cost of construction. Indices have been established for material, installation, and total cost, broken down by building system. This allows the estimator to analyze specific project elements, such as a subcontractor package, as well as look at material purchasing costs and the cost of labor relative to a particular location. In this table the national average city would have a total index equal to 100. In 1995 the most expensive city was New York City with an index of 133.8, and the least expensive city was Columbia, South Carolina, with an index of 77.9. The city closest to the national average was Spokane, Washington, with an index of 99.9. The material index for New York is 111.2, indicating that the cost of materials there is 11.2 percent above the national average. The installation (labor) index is 158.1, or 58.1 percent above the national average. In Columbia the material index is only 4.7 percent below the national average, but the installation costs are extremely low with an index of 59.2; that is, 40.8 percent below the national average.

However, many projects are built in areas without a readily available city cost index. In that situation the estimator must analyze the project and create the proper adjustment. Often the easiest way is to look at the cities that have indices and select the location most similar with respect to installation and material costs. In some cases different cities may be selected for material and labor, with the average of the two then taken. Companies that do a lot of work in the same location usually develop their own location index.

Square Foot Project Size Modifier

One factor that affects the S.F. cost of a particular building is the size. In general, for buildings built to the same specifications in the same locality, the larger building will have the lower S.F. Cost. This is due mainly to the decreasing contribution of the exterior walls plus the economy of scale usually achievable in larger buildings. The Area Conversion Scale shown below will give a factor to convert costs for the typical size building to an adjusted cost for the particular project.

The Square Foot Base Size lists the median costs, most typical project size in our accumulated data and the range in size of the projects.

The Size Factor for your project is determined by dividing your project area in S.F. by the typical project size for the particular Building Type. With this factor, enter the Area Conversion Scale at the appropriate Size Factor and determine the appropriate cost multiplier for your building size.

Example: Determine the cost per S.F. for a 100,000 S.F. Mid-rise apartment building.

$$\frac{\text{Proposed building area} = 100,000 \text{ S.F.}}{\text{Typical size from below} = 50,000 \text{ S.F.}} = 2.00$$

Enter Area Conversion scale at 2.0, intersect curve, read horizontally the appropriate cost multiplier of .94. Size adjusted cost becomes .94 x $62.35 = $58.60 based on national average costs.

Note: For Size Factors less than .50, the Cost Multiplier is 1.1
For Size Factors greater than 3.5, the Cost Multiplier is .90

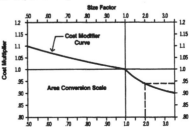

Square Foot Base Size							
Building Type	Median Cost per S.F.	Typical Size Gross S.F.	Typical Range Gross S.F.	Building Type	Median Cost per S.F.	Typical Size Gross S.F.	Typical Range Gross S.F.
Apartments, Low Rise	$ 49.45	21,000	9,700 - 37,200	Jails	$144.00	13,700	7,500 - 28,000
Apartments, Mid Rise	62.35	50,000	32,000 - 100,000	Libraries	87.70	12,000	7,000 - 31,000
Apartments, High Rise	71.25	310,000	100,000 - 650,000	Medical Clinics	84.15	7,200	4,200 - 15,700
Auditoriums	83.25	25,000	7,600 - 39,000	Medical Offices	79.60	6,000	4,000 - 15,000
Auto Sales	49.55	20,000	10,800 - 28,600	Motels	60.60	27,000	15,800 - 51,000
Banks	111.00	4,200	2,500 - 7,500	Nursing Homes	82.65	23,000	15,000 - 37,000
Churches	74.75	9,000	5,300 - 13,200	Offices, Low Rise	66.10	8,600	4,700 - 19,000
Clubs, Country	74.10	6,500	4,500 - 15,000	Offices, Mid Rise	70.35	52,000	31,300 - 83,100
Clubs, Social	73.50	10,000	6,000 - 13,500	Offices, High Rise	88.60	260,000	151,000 - 468,000
Clubs, YMCA	75.80	28,300	12,800 - 39,400	Police Stations	111.00	10,500	4,000 - 19,000
Colleges (Class)	96.70	50,000	23,500 - 98,500	Post Offices	82.25	12,400	6,800 - 30,000
Colleges (Science Lab)	128.00	45,600	16,600 - 80,000	Power Plants	626.00	7,500	1,000 - 20,000
College (Student Union)	107.00	33,400	16,000 - 85,000	Religious Education	64.15	9,000	6,000 - 12,000
Community Center	77.85	9,400	5,300 - 16,700	Research	114.00	19,000	6,300 - 45,000
Court Houses	104.00	32,400	17,800 - 106,000	Restaurants	101.00	4,400	2,800 - 6,000
Dept. Stores	46.00	90,000	44,000 - 122,000	Retail Stores	48.80	7,200	4,000 - 17,600
Dormitories, Low Rise	74.15	24,500	13,400 - 40,000	Schools, Elementary	71.10	41,000	24,500 - 55,000
Dormitories, Mid Rise	95.65	55,600	36,100 - 90,000	Schools, Jr. High	72.45	92,000	52,000 - 119,000
Factories	44.35	26,400	12,900 - 50,000	Schools, Sr. High	73.00	101,000	50,500 - 175,000
Fire Stations	79.25	5,800	4,000 - 8,700	Schools, Vocational	72.20	37,000	20,500 - 82,000
Fraternity Houses	67.95	12,500	8,200 - 14,800	Sports Arenas	56.95	15,000	5,000 - 40,000
Funeral Homes	77.95	7,800	4,500 - 11,000	Supermarkets	48.90	20,000	12,000 - 30,000
Garages, Commercial	55.30	9,300	5,000 - 13,600	Swimming Pools	90.25	13,000	7,800 - 22,000
Garages, Municipal	67.35	8,300	4,500 - 12,600	Telephone Exchange	132.00	4,500	1,200 - 10,600
Garages, Parking	27.15	163,000	76,400 - 225,300	Terminals, Bus	64.85	11,400	6,300 - 16,500
Gymnasiums	69.30	19,200	11,600 - 41,000	Theaters	72.95	10,500	8,800 - 17,500
Hospitals	135.00	55,000	27,200 - 125,000	Town Halls	78.40	10,800	4,800 - 23,400
House (Elderly)	67.40	37,000	21,000 - 66,000	Warehouses	32.55	25,000	8,000 - 72,000
Housing (Public)	61.60	36,000	14,400 - 74,400	Warehouse & Office	37.00	25,000	8,000 - 72,000
Ice Rinks	70.40	29,000	27,200 - 33,600				

Figure 5–3

Determining the project size modifier.

From *Means Assemblies Cost Data 1995*. Copyright R. S. Means Co., Inc., Kingston, MA, 617-585-7880, all rights reserved.

CITY COST INDEXES   R13.3-010   Building Systems

DIV. NO.	BUILDING SYSTEMS	NEW YORK														
		ALBANY			BINGHAMTON			BUFFALO			NEW YORK			ROCHESTER		
		MAT.	INST.	TOTAL	MAT.	INST.	TOTAL	MAT.	INST.	TOTAL	MAT.	INST.	TOTAL	MAT.	INST.	TOTAL
1-2	FOUND/SUBSTRUCTURES	88.4	101.1	96.6	105.6	87.4	93.9	104.4	113.1	110.0	126.7	154.9	144.8	96.7	105.5	102.4
3	SUPERSTRUCTURES	97.2	109.3	102.7	97.8	106.4	101.7	100.0	109.3	104.2	115.3	153.2	132.5	100.8	111.8	105.8
4	EXTERIOR CLOSURE	106.5	97.1	102.0	104.4	90.4	97.7	102.7	120.0	110.9	128.8	160.9	144.1	109.5	103.0	106.4
5	ROOFING	90.9	96.5	93.4	100.4	86.7	94.4	100.1	111.3	105.0	105.5	150.2	125.0	95.4	103.8	99.1
6	INTERIOR CONSTRUCTION	97.4	90.5	94.6	94.4	79.0	88.2	93.7	120.2	104.5	105.1	161.8	128.2	97.1	102.0	99.1
7	CONVEYING	100.0	99.2	99.7	100.0	102.2	100.6	100.0	112.3	103.5	100.0	145.7	113.1	100.0	103.3	100.9
8	MECHANICAL	100.0	91.2	96.2	100.3	80.4	91.7	99.8	103.9	101.6	100.0	165.6	128.4	99.8	93.4	97.0
9	ELECTRICAL	106.6	94.0	98.2	100.9	87.0	91.7	102.0	103.7	103.2	117.4	163.5	148.0	107.7	93.6	98.3
11	EQUIPMENT	100.0	89.8	99.3	100.0	74.3	98.2	100.0	119.5	101.3	100.0	166.5	104.4	100.0	100.3	100.0
12	SITE WORK	72.2	110.6	100.9	98.7	90.9	92.8	97.4	93.8	94.7	142.1	136.0	137.5	72.6	110.2	100.7
1-12	WEIGHTED AVERAGE	98.8	97.9	98.4	99.4	88.9	94.3	99.3	109.5	104.2	111.2	158.1	133.8	100.4	101.7	101.0

DIV. NO.	BUILDING SYSTEMS	PENNSYLVANIA						RHODE ISLAND			SOUTH CAROLINA					
		READING			SCRANTON			PROVIDENCE			CHARLESTON			COLUMBIA		
		MAT.	INST.	TOTAL	MAT.	INST.	TOTAL	MAT.	INST.	TOTAL	MAT.	INST.	TOTAL	MAT.	INST.	TOTAL
1-2	FOUND/SUBSTRUCTURES	92.3	96.6	95.0	98.7	97.6	98.0	98.9	109.2	105.5	85.9	65.1	72.5	83.7	64.9	71.6
3	SUPERSTRUCTURES	95.4	109.4	101.7	99.5	108.4	103.5	104.5	108.7	106.4	93.5	73.0	84.2	93.4	73.1	84.2
4	EXTERIOR CLOSURE	107.4	89.0	98.7	106.9	95.4	101.5	119.8	112.8	116.5	95.8	47.5	72.8	95.3	47.6	72.6
5	ROOFING	99.2	109.5	103.7	99.0	94.3	96.9	98.8	105.9	101.9	92.5	55.7	76.4	92.6	56.2	76.8
6	INTERIOR CONSTRUCTION	95.8	88.8	92.9	96.4	90.8	94.1	96.6	111.9	102.8	93.1	51.3	76.1	93.1	54.1	77.2
7	CONVEYING	100.0	101.6	100.4	100.0	104.1	101.2	100.0	114.0	104.0	100.0	75.5	92.9	100.0	75.5	92.9
8	MECHANICAL	100.2	96.8	98.7	100.0	92.4	96.7	99.8	103.2	101.3	99.9	61.8	83.4	99.9	52.1	79.2
9	ELECTRICAL	104.3	89.2	94.3	106.6	88.4	94.5	99.7	110.3	106.8	96.2	48.0	64.2	96.2	50.7	65.9
11	EQUIPMENT	100.0	86.7	99.1	100.0	88.0	99.2	100.0	104.3	100.2	100.0	52.7	96.8	100.0	57.3	97.1
12	SITE WORK	117.0	108.8	110.9	96.5	110.3	106.8	97.4	104.8	102.9	90.2	82.4	84.4	89.5	82.4	84.2
1-12	WEIGHTED AVERAGE	99.6	96.9	98.3	100.4	96.7	98.6	102.4	108.6	105.4	95.5	60.2	78.4	95.3	59.2	77.9

DIV. NO.	BUILDING SYSTEMS	WASHINGTON						WEST VIRGINIA						WISCONSIN		
		SPOKANE			TACOMA			CHARLESTON			HUNTINGTON			GREEN BAY		
		MAT.	INST.	TOTAL	MAT.	INST.	TOTAL	MAT.	INST.	TOTAL	MAT.	INST.	TOTAL	MAT.	INST.	TOTAL
1-2	FOUND/SUBSTRUCTURES	106.7	91.3	96.8	93.1	110.8	104.5	96.2	84.4	88.6	99.8	89.1	92.9	100.4	89.8	93.6
3	SUPERSTRUCTURES	98.2	89.0	94.0	101.1	96.2	98.9	96.8	92.3	94.8	97.0	97.6	97.3	97.6	91.0	94.6
4	EXTERIOR CLOSURE	121.0	89.9	106.3	132.4	100.5	117.3	95.8	82.6	89.5	96.3	84.1	90.5	96.5	81.3	89.3
5	ROOFING	169.5	84.5	132.4	106.7	96.7	102.3	93.0	82.4	88.4	92.5	86.5	89.8	98.1	80.0	90.2
6	INTERIOR CONSTRUCTION	132.4	86.6	113.7	121.3	99.6	112.5	93.6	84.4	89.9	92.9	87.5	90.7	106.2	81.3	96.1
7	CONVEYING	100.0	99.3	99.8	100.0	107.4	102.1	100.0	99.0	99.7	100.0	85.0	95.6	100.0	92.3	97.7
8	MECHANICAL	100.5	91.8	96.7	100.1	107.3	103.2	99.9	78.3	90.5	99.9	80.2	91.4	100.1	83.6	93.0
9	ELECTRICAL	90.9	81.1	84.4	101.6	90.8	94.4	96.2	83.1	87.5	96.2	85.7	89.2	92.5	84.6	87.2
11	EQUIPMENT	100.0	86.5	99.1	100.0	104.5	100.3	100.0	85.2	99.0	100.0	88.4	99.2	100.0	81.6	98.7
12	SITE WORK	95.2	93.5	93.9	89.1	125.4	116.3	103.6	85.3	89.9	106.2	85.2	90.4	78.5	96.0	91.6
1-12	WEIGHTED AVERAGE	110.4	88.6	99.9	108.2	102.3	105.3	97.0	84.5	90.9	97.2	87.0	92.2	99.2	86.0	92.8

Figure 5–4
City cost indices for selected cities.
From *Means Assemblies Cost Data 1995*. Copyright R. S. Means Co., Inc., Kingston, MA, 617-585-7880, all rights reserved.

To adjust our motel project to the three cities above, a ratio could be set up as follows:

$$\frac{\text{Estimated cost}}{100} = \frac{\text{Adjusted cost for city}}{\text{City index}}$$

New York City

$$\frac{2,014,650}{100} = \frac{\text{Adjusted New York cost}}{133.8}$$

New York City cost = \$2,695,601

Columbia

$$\frac{2,014,650}{100} = \frac{\text{Adjusted Columbia cost}}{77.9}$$

Columbia cost = $1,569,412

Spokane

$$\frac{2,014,650}{100} = \frac{\text{Adjusted Spokane cost}}{99.9}$$

Spokane cost = $2,012,635

This example illustrates the importance of location to the cost of construction. As can be seen, the cost of constructing the motel in New York is more than $1 million over the estimated cost in Columbia. Unadjusted prices essentially give you the cost of construction in Spokane, Washington.

The last adjustment that will be made for this estimate is for time. This estimate was prepared using 1995 data, so that the estimate to this point is valid for a construction start in early 1995. A project set to begin in early 1997 would have to be adjusted for expected increases in labor and material. Indices for past projects can be adjusted to the present by comparing actual past project costs to the index of that year as compared to the index today.

Assuming a 1980 project cost of $2 million, calculation of the same project cost in 1995 would be figured as follows:

$$\frac{\text{Past project cost}}{\text{Index past year}} = \frac{\text{Current project cost}}{\text{Index 1995}}$$

$$\frac{2,000,000}{59.5} = \frac{\text{Current project cost}}{100}$$

Current project cost = $3,361,344

The difficulty often faced in conducting estimates is that indices are not available for future years, so past and current trends must be looked at and projected to the future. *Note*: Means will often suggest in the Historical Index table an escalation rate for future years.

Another source of cost index information is *Engineering News-Record* (ENR), which publishes on a weekly basis cost indices dating back to 1913, the year which has been set as the baseline. Each week the magazine looks at different materials, industries, regions, and the like and makes comparisons as well as projections. Indices are published for Construction Cost, Building Cost, Common Labor, Skilled Labor, and Materials. *ENR* indices can be used just as the Means indices to adjust a project for time.

To construct a roadway in August 1994 that cost $5 million in August 1981, the adjustment would be figured as follows:

ENR Market Trends

Latest Week

COST INDEXES ENR 20-cities 1913 = 100	July 30* index value	Change from last month %	Change from last year %
Construction Cost	3,575.24	+ 0.4	+ 8.2
Building Cost	2,117.84	+ 0.4	+ 7.3
Common labor (CC)	6,890.13	+ 0.5	+ 8.9
Skilled labor (BC)	3,058.80	+ 0.8	+ 8.1
Materials	1,543.54	0	+ 6.3

*Official August Indexes

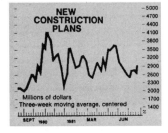

NEW CONSTRUCTION PLANS

Millions of dollars
Three-week moving average, centered

SEPT 1980 1981 MAR JUN

50-state totals ENR-reported $ millions	Week of Aug. 6 1981	Cum. 32 weeks change '80-'81 %
BIDDING VOLUME		
Total construction*	1,216.4	46,284.4	+ 10
Heavy & highway	356.9	16,728.0	– 1
Nonresidential bldg.	641.6	23,573.6	+ 17
Housing, multiunit	217.8	5,982.8	+ 20
NEW PLANS		
Total construction*	3,870.0	96,740.8	+ 9
Heavy & highway, total.	1,018.3	26,807.7	– 12
Water use & control	478.3	9,905.7	+ 12
Waterworks	55.3	2,508.2	+ 20
Sewerage	321.4	4,901.0	+ 4
Treatment plants	183.6	2,598.9	+ 13
Earthwork, waterways	101.6	2,496.6	+ 23
Transportation	416.5	9,305.2	– 45
Highways	212.1	5,163.2	– 55
Bridges	43.5	1,133.3	– 60
Airports	74.5	2,142.8	+ 16
Terminals, hangars	48.5	1,264.7	0
Elec, gas, comm	49.5	5,721.0	+ 77
Other heavy const	73.0	1,692.9	+ 9
Nonresidential bldg	2,122.5	51,124.0	+ 19
Manufacturing	249.8	12,741.9	+ 58
Commercial	1,074.7	18,866.6	+ 19
Offices	792.3	11,820.2	+ 31
Stores, shopping ctrs.	142.4	4,630.2	– 3
Educational	108.7	4,098.8	+ 4
College, university	54.5	1,477.2	– 5
Medical	408.2	5,967.9	+ 31
Hospital	322.0	4,687.0	+ 27
Other	281.2	9,448.9	– 11
Housing, multiunit*	729.1	18,809.2	+ 21
Apartments	492.1	13,449.8	+ 17

*Excludes 1-2 family houses. Minimum sizes included are: Industrial plants, heavy and highway construction, $100,000; buildings, $500,000.

NEW CONSTRUCTION CAPITAL

	Week of July 30 $ millions	Cum. 31 week change '80-'81 %
Total new capital	826.3	22,166.4	+22
Corporate securities	500.0	4,233.5	+64
State and municipal	326.3	17,932.7	+15
Housing	0	8,084.4	– 4
Other bldg and heavy	326.3	9,848.5	+36

Latest Month

WAGE RATES, 20 cities' average	Aug. 1981	July 1981	%change from July 1981	%change from Aug. 1980
Common	13.09		+ 0.5	+ 8.9
Skilled (average 3 trades)	16.98		+ 0.8	+ 8.2
Bricklayers	16.77		+ 0.3	+ 9.2
Structural ironworkers	17.66		+ 1.7	+ 9.3
Carpenters	16.50		+ 0.3	+ 5.8

MATERIAL PRICES, 20-cities' average

	Aug. 1981	%change from July 1981	%change from Aug. 1980
Structural steel (average mill), base, per cwt	21.65	0	+ 13.8
Lumber, 2x4 fir, per Mbf, CL	335.83	– 0.1	+ 1.6
Lumber, 2x4 pine, per Mbf, CL	305.93	+ 0.3	– 2.0
Cement, bulk, per ton, TL†	60.19	– 0.1	+ 1.6
Sand, per ton, CL†	6.10	0	+ 6.8
Ready-mix concrete, 3,000 psi, per cu yd, 15 cu yd	45.16	+ 0.5	+ 8.3
Crushed stone, 1½'', per ton, CL†	7.50	0	+ 10.9
Concrete blocks, sand/gravel, 8''x8''x16'', ea, TL*	0.68	0	+ 3.0

	July 1981	%change from June 1981	%change from July 1980
Gypsum sheathing, ½'x2'x8', per Msf, TL*	147.21	+ 0.2	+ 10.6
Plywood, plyform, ¾'', per Msf, CL†	608.80	+ 0.2	+ 10.5
Plywood, ⅝, per Msf, CL†	358.39	+ 0.3	+ 5.8
Brick, common, per M, TL*	162.96	– 0.1	+ 7.5
Reinforced bars, whse base, per cwt	18.82	0	– 2.1
Asphalt paving, AC-20	173.96	+ 13.0	+ 40.1
Vitrified clay sewer pipe, premium joint, 12'', per ft, CL*	5.74	+ 0.7	+ 8.3
Concrete sewer pipe, 12'', premium joint, per ft, CL*	6.49	– 0.3	+ 4.3

*Delivered ††t.o.b. city CL-Carlots TL-Trucklots

ENR INDEX REVIEW

Base year = 100	Construction Cost 1913	Construction Cost 1967	Building Cost 1913	Building Cost 1967	Wage Rates Skilled 1913	Wage Rates Skilled 1967	Wage Rates Common 1913	Wage Rates Common 1967
1980								
Aug.	3304.20	307.61	1973.65	292.13	2829	284	6327	312
Sept.	3318.81	308.97	1975.53	292.41	2845	285	6376	313
Oct.	3326.82	309.71	1975.97	292.48	2867	287	6418	315
Nov.	3356.57	312.48	2000.09	296.05	2871	288	6437	316
Dec.	3376.13	314.33	2017.40	298.61	2890	290	6463	317
1980 aver.	**3237.28**	**301.44**	**1943.27**	**287.73**	**2767**	**278**	**6168**	**304**
1981								
Jan.	3372.02	313.92	2014.93	298.25	2895	291	6463	319
Feb.	3373.19	314.03	2016.10	298.42	2895	291	6463	319
Mar.	3383.71	315.01	2013.64	298.05	2902	291	6504	321
Apr.	3452.05	321.37	2064.08	305.52	2906	292	6555	324
May	3472.92	323.31	2076.06	307.29	2936	295	6608	327
June	3510.28	326.79	2082.78	308.29	2970	299	6723	333
July	3562.32	331.64	2108.62	312.11	3035	305	6857	339
Aug.	3575.24	332.84	2117.84	313.48	3059	307	6890	340

Trends to Watch

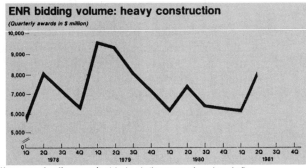

ENR bidding volume: heavy construction
(Quarterly awards in $ million)

Heavy construction awards picked up in the quarter after a long decline.

Figure 5–5
ENR Market Trends, August 6, 1981.

ENR MARKET TRENDS

Latest Week

COST INDEXES ENR 20 cities 1913=100	Aug. index value	Change from last month %	year %
Construction Cost	5432.95	+0.4	+ 3.9
Building Cost	3109.43	+0.1	+ 3.2
Common labor (CC)	10943.68	+0.7	+ 3.8
Skilled labor (BC)	4835.41	+0.4	+ 2.4
Materials	2055.41	−0.4	+ 4.3

Construction contract awards
$ mil.
Four-week moving average
5200 4800 4400 4000 3600 3200 2800 2400 2000 1600 1200 800 400 0
N D J F M A M J J A
1993 1994

National totals ENR-reported $ mil.	Week ending Aug. 1	6 mos. cum. 1994	% chg. '93-94
CONTRACT AWARDS			
Total construction¹	4,432.2	80,395.1	+ 4
Heavy & highway	1,589.1	31,030.6	0
Water use & control	463.6	8,042.5	− 3
Waterworks	171.7	2,739.4	− 2
Sewerage	175.6	3,384.5	− 8
Dams, waterway dev.	116.3	1,918.5	+ 6
Transportation	814.3	16,795.7	+ 12
Highways	577.4	11,842.5	+ 17
Bridges & tunnels	175.8	3,929.4	+ 8
Airports, incl. buildings	61.1	1,023.8	− 19
Electric, gas, communic.	36.1	1,655.9	− 52
Military, space	1.0	4.0	+ 5
Other heavy const.	274.0	4,532.4	+ 6
Total nonres. bldgs.	2,533.5	42,484.1	+ 5
Manufacturing bldgs.	445.9	3,707.6	+ 8
Commercial buildings	803.4	16,708.7	+ 10
Offices, banks	233.5	6,055.4	+ 5
Stores, shopping ctrs	267.8	6,130.8	+ 17
Other comml. service	302.1	4,522.4	+ 8
Government buildings	239.9	2,389.3	+ 23
Administration	41.1	551.4	0
Post offices	20.7	79.3	+ 24
Prisons	146.4	1,281.7	+ 38
Police, fire	31.7	476.9	+ 22
Educational buildings	601.7	9,436.1	0
Primary	437.9	6,230.4	0
College	79.5	1,595.0	+ 30
Laboratories	84.3	1,610.7	− 20
Medical buildings	176.7	4,929.7	− 1
Hospitals	87.0	2,827.8	− 2
Nursing	89.7	2,102.0	0
Other nonresidential	265.9	5,312.6	− 5
Total multiunit hsg.²	309.6	6,880.5	+ 23
Apartments	242.3	4,975.6	+ 13
Hotels, motels, dorms	67.3	1,904.9	+ 59

Source: ENR-F.W. Dodge Division
¹Minimum size: Contract awards, $50,000
²Excludes 1-2 family houses

NEW PUBLIC CONSTRUCTION CAPITAL $ mil.	Week of Aug. 15	30 weeks cum. Latest figure	Annual % chg.
State and municipal	1,085.3	25,773.5	− 38
Housing	137.8	1,823.1	− 55
Other bldg. and heavy	947.5	23,950.4	− 41

Latest Month

NEW CONSTRUCTION CONTRACTS
1987=100

	March	April	% chg. mo. ago	% chg. yr. ago
ENR total contract awards	100	88	−12.0	− 4.3
Nonresidential/multiunit	88	74	−15.9	− 5.1
Heavy and highway	132	126	− 4.5	− 2.3

Source: ENR-F.W. Dodge Division. Shows seasonally adjusted annual rate.

STEEL SHIPMENTS

	March	3 mos.	mo. ago	Cum. '93-94
All construction	722	1,963	− 4	+ 9
Contractor	267	691	− 2	+ 15
Structural & piling	431	1,323	− 7	+ 4

Source: DRI Steel Service. Shipment figures are in thousands of net tons.

CONSTRUCTION MACHINERY DISTRIBUTOR INDEXES
1988=100

	May	% chg. mo. ago	% chg. yr. ago	5-mos. avg.	Cum. % chg. '93-94
Sales—AED survey of 134 dealers	120	+ 1	+ 20	120	+ 20
Inventories—AED survey of 134 dealers	111	+ 1	+ 11	112	+ 12
Inventories—Sales ratio in month	4:4	0	− 7	4:7	− 7

Source: Associated Equipment Distributors

STATE AND MUNICIPAL BOND SALES FOR CONSTRUCTION

	May* 1994	% chg. April '94	1994 cum. $ mil.	% chg. yr. ago
Total bond sales (ENR-reported)	3,943.9	+ 21	18,282.0	− 37
Buildings, total	1,618.9	− 10	8,568.4	− 20
Schools	560.1	− 13	3,239.3	− 20
Housing	468.4	+ 43	1,363.2	+ 82
Other buildings	590.4	− 28	3,965.9	− 32
Heavy and highway construction, total	1,208.9	+ 25	5,752.3	− 47
Waterworks	139.3	− 44	976.9	+ 2
Sewerage, pollution control	415.7	+ 45	1,591.3	+ 11
Bridges	0.0	0	660.1	− 55
Highways	0.7	− 56	456.4	− 53
Earthwork, irrigation, drainage	0.0	0	0.0	−100
Electric and gas utilities	193.1	+410	1,004.7	− 79
Airports	112.0	− 71	698.7	+ 47
Mass transit	348.1	+..	364.2	− 60
General improvement and unclassified	1,116.1	+121	3,961.3	− 46

*Based on average week

CONTRACTOR FAILURES

	Numbers 12 mos. 1992	12 mos. 1993†	% chg. '92-93	Liabilities* 12 mos. 1992	12 mos. 1993†	% chg. '92-93
All contractors	12,452	10,388	− 17	5,036.8	2,133.9	− 58
General building contractors	4,659	3,864	− 17	2,868.3	1,312.7	− 54
Other general contractors	598	505	− 16	152.1	93.5	− 39
Special trade contractors	7,195	6,019	− 16	2,016.4	727.7	− 64

Source: The Dun & Bradstreet Corp.; *$ mil.; †preliminary

Trends to Watch

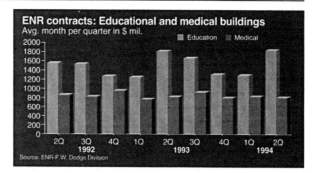

ENR contracts: Educational and medical buildings
Avg. month per quarter in $ mil.
■ Education ■ Medical
2000 1800 1600 1400 1200 1000 800 600 400 200 0
2Q 3Q 4Q | 1Q 2Q 3Q 4Q | 1Q 2Q
1992 | 1993 | 1994
Source: ENR-F.W. Dodge Division

Figure 5–6
ENR Market Trends, August 22, 1994.

Construction Cost Index (*ENR*) Aug. 1981 = 3,575 (See Fig. 5–5)

Construction Cost Index (*ENR*) Aug. 1994 = 5,433 (See Fig. 5–6)

$$\frac{\$5,000,000}{3,575} = \frac{\text{Construction cost 1994}}{5,433}$$

Construction cost 1994 = $7,598,601

Both the Means and *ENR* indices can be used to adjust projects for time. However, since *ENR* uses a baseline of 1913 = 100 and Means a baseline of 1975 = 100, the tables cannot be mixed.

Presentation

For the purposes of discussion, assume that the motel we have been estimating will be built in New York City and construction will start in 1997. Assuming an increase of 2.5 percent per year the 1995 New York price of $2,695,601 would increase 5 percent to $2,830,381. This price would reflect a 50-unit, high quality motel built in New York City with the major construction occurring in 1997. The price is based on R.S. Means 1995 data.

In presenting any estimate it is important to consider the purpose of the estimate as well as understand what is included and not included in the price. It is also important to understand all of the underlying assumptions and the accuracy of the data. All of this must be integral to the presentation to the owner. Conceptual estimates are often the first costs that are put before the owner. Although they are normally accomplished with little information, they tend to be the number most remembered. It is therefore important for the estimator to qualify the price, that is, list all assumptions made and identify just what information has been considered and what information must still be researched and adjusted for.

In the motel example the cost of land, demolition if required, and design fees have not been included in the price and would have to be added. The estimate, which took less than an hour to complete, would not have considered any unusual design features, special code requirements, or a high level of site work.

Square Foot Estimating

Approach

The method of assembling a square foot estimate is similar to that used for a conceptual estimate except more information is required and costs are tabulated per square foot, not by service unit. The estimate is still used primarily by the owner for budgeting purposes and is still conducted during the conceptual stage of the project. Cost data can be taken from outside data sources or can be developed by the designer, owner, or construction company. Adjustments are made the same as in the ROM estimate, and presentation considerations are similar. These estimates take slightly longer than a ROM estimate and should provide an accuracy in the ±15 percent range.

Data

Companies that have information for ROM estimates should also have the data needed for square foot estimates. If information is not available in-house, Means has two square foot data books, each published utilizing different source material. The Assemblies Cost Data book (see Fig. 5–7) bases its costs on actual completed projects as reported by owners, designers, and contractors. This is the same source of information used in the conceptual estimate, only broken down on a square foot basis. It uses the same ¼, median, and ¾ quality distinction and is based on 11,500 projects. The second source is the Square Foot Data book (see Figs. 5–8 and 5–9), which is based on stereotypical models that have been created for different project types. This allows Means to determine the quantities of materials and labor required to build each project at the sizes tabulated. With the quantities held constant, each year the unit prices are adjusted to reflect current costs for labor and materials.

Each square foot approach has its own advantages and disadvantages. The reported square foot costs, in the Assemblies Cost book, are derived from actual projects that were built around the country; they have been localized to the national average. These are true costs reflecting actual contractor costs including overhead and profit. The disadvantage of making a square foot estimate using that reference is that specific project conditions are not known, thereby making it difficult to know exactly what the square foot cost includes. The modeled data from the Square Foot book is more specific. It identifies the costs associated with different combinations of structural and exterior closure. It also allows specific pricing for common project additives such as basements, and adjustments for story height and building perimeter. The perimeter adjustment factor allows the estimator to properly figure the added cost associated with unusually shaped buildings. This data book also provides a detailed breakout of the model for the project highlighted on the facing page (see Fig. 5–9). The detail breakout allows the estimator to make small adjustments for items in the model by deleting and adding line items as necessary.

Using the Reported Square Foot Data

What is the cost of a 15,000 sq.ft. library? Assume median quality and 1995 data (see again Fig. 5–7).

size in sq.ft. × cost per sq.ft. = total cost
15,000 sq.ft. × $87.70/sq.ft. = $1,315,500

		14.1 000	S.F. & C.F. Costs	UNIT	UNIT COSTS			% OF TOTAL			
					1/4	MEDIAN	3/4	1/4	MEDIAN	3/4	
500	0100	Site work		S.F.	5.35	7.80	13.05	8%	11.60%	16.10%	500
	1800	Equipment			1.20	1.97	3.23	2.10%	2.90%	4.60%	
	2720	Plumbing			3.04	4.27	5.40	6.80%	9%	11.50%	
	2730	Heating, ventilating, air conditioning			1.62	3.15	3.36	4.20%	6%	6.40%	
	2900	Electrical			2.70	3.88	5.55	5%	6.50%	8.10%	
	3100	Total: Mechanical & Electrical			8.15	11.75	16.15	15.60%	19.20%	23.50%	
	9000	Per apartment, total cost		Apt.	47,600	53,000	67,100				
	9500	Total: Mechanical & Electrical		*	7,800	10,700	13,400				
510	0010	ICE SKATING RINKS		S.F.	40.30	70.40	97.50				510
	0020	Total project costs		C.F.	2.81	2.88	3.32				
	2720	Plumbing		S.F.	1.43	2.27	2.68	3.10%	3.20%	5.60%	
	2900	Electrical			4.10	4.52	6.30	6.70%	10.10%	15%	
	3100	Total: Mechanical & Electrical			7.25	10.45	13.05	9.90%	25.90%	29.80%	
520	0010	JAILS		S.F.	115	144	192				520
	0020	Total project costs		C.F.	10.70	12.70	17.20				
	1800	Equipment		S.F.	4.87	12.95	22.50	4%	8.90%	15.10%	
	2720	Plumbing			11.45	15.10	18.35	7%	8.90%	13.30%	
	2770	Heating, ventilating, air conditioning			10.50	14.30	26.50	7.40%	9.40%	17.70%	
	2900	Electrical			12.80	16.60	20.40	7.90%	11.30%	12.40%	
	3100	Total: Mechanical & Electrical			30	43.30	57.15	24.30%	30.10%	31.70%	
530	0010	LIBRARIES		S.F.	69.55	87.70	109				530
	0020	Total project costs		C.F.	4.96	6	7.55				
	0500	Masonry		S.F.	3.91	7.15	13.75	4.50%	7.40%	10.70%	
	1800	Equipment			1	2.60	4.31	1.10%	2.80%	4.80%	
	2720	Plumbing			2.80	3.93	5.35	3.40%	4.60%	5.60%	
	2770	Heating, ventilating, air conditioning			6.20	10.20	13.25	8.70%	11%	14.60%	
	2900	Electrical			7.15	9.30	11.90	8.20%	10.90%	12.10%	
	3100	Total: Mechanical & Electrical			14.85	19.55	28.75	17.10%	22%	28%	
550	0010	MEDICAL CLINICS		S.F.	68.10	84.15	106				550
	0020	Total project costs		C.F.	5.20	6.70	8.85				
	1800	Equipment		S.F.	1.82	3.88	5.85	1.80%	4.80%	6.80%	
	2720	Plumbing			4.71	6.50	8.70	6.10%	8.40%	10%	
	2770	Heating, ventilating, air conditioning			5.60	7.20	10.60	6.60%	9.20%	11.30%	
	2900	Electrical			5.95	8.30	11	8.10%	10%	11.90%	
	3100	Total: Mechanical & Electrical			14.40	18.35	25.80	18.60%	23.60%	29.60%	
	3500										
570	0010	MEDICAL OFFICES		S.F.	63.35	79.60	98.20				570
	0020	Total project costs		C.F.	4.87	6.65	8.95				
	1800	Equipment		S.F.	2.15	4.26	6.05	3%	5.80%	7.10%	
	2720	Plumbing			3.69	5.65	7.65	5.60%	6.80%	8.60%	
	2770	Heating, ventilating, air conditioning			4.35	6.50	8.30	6.10%	8%	9.50%	
	2900	Electrical			5.15	7.55	10.50	7.60%	9.70%	11.60%	
	3100	Total: Mechanical & Electrical			11.95	16.70	22.35	16.60%	21.20%	26%	
590	0010	MOTELS		S.F.	41.80	60.60	76.15				590
	0020	Total project costs		C.F.	3.59	5.06	8.30				
	2720	Plumbing		S.F.	4.17	5.20	6.35	9.40%	10.60%	12.50%	
	2770	Heating, ventilating, air conditioning			2.20	3.79	5.55	4.90%	5.60%	8.20%	
	2900	Electrical			3.88	4.90	6.35	7.10%	8.20%	10.40%	
	3100	Total: Mechanical & Electrical			9.40	12.05	16.60	18.50%	23.10%	26.20%	
	5000										
	9000	Per rental unit, total cost		Unit	21,000	31,500	40,700				
	9500	Total: Mechanical & Electrical			4,025	5,700	6,150				

SQUARE FOOT

Figure 5–7

Sample costs per square foot for library construction.

Costs per square foot of floor area

EXTERIOR WALL	S.F. Area	7000	10000	13000	16000	19000	22000	25000	28000	31000
	L.F. Perimeter	240	300	336	386	411	435	472	510	524
Face Brick with Concrete Block Back-up	R/Conc. Frame	98.70	93.45	89.00	87.00	84.50	82.60	81.65	80.90	79.65
	Steel Frame	95.20	89.90	85.50	83.45	80.95	79.10	78.15	77.40	76.15
Limestone with Concrete Block	R/Conc. Frame	110.85	104.10	98.20	95.55	92.15	89.65	88.35	87.40	85.65
	Steel Frame	107.35	100.60	94.65	92.05	88.65	86.10	84.85	83.85	82.15
Precast Concrete Panels	R/Conc. Frame	94.15	89.45	85.55	83.80	81.60	80.00	79.15	78.50	77.40
	Steel Frame	90.60	85.90	82.05	80.25	78.10	76.45	75.60	75.00	73.90
Perimeter Adj., Add or Deduct	Per 100 L.F.	12.35	8.65	6.65	5.40	4.55	3.95	3.45	3.10	2.80
Story Hgt. Adj., Add or Deduct	Per 1 Ft.	1.90	1.65	1.40	1.30	1.15	1.05	1.05	1.00	.90
For Basement, add $24.95 per square foot of basement area										

The above costs were calculated using the basic specifications shown on the facing page. These costs should be adjusted where necessary for design alternatives and owner's requirements. Reported completed project costs, for this type of structure, range from $52.80 to $136.75 per S.F.

Common additives

Description	Unit	$ Cost
CARRELS Hardwood	Each	610 - 850
CLOSED CIRCUIT SURVEILLANCE, One station		
Camera and monitor	Each	1300
For additional camera stations, add	Each	690
ELEVATORS, Hydraulic passenger, 2 stops		
1500# capacity	Each	45,100
2500# capacity	Each	46,400
3500# capacity	Each	50,100
EMERGENCY LIGHTING, 25 watt, battery operated		
Lead battery	Each	370
Nickel cadmium	Each	610
FLAGPOLES, Complete		
Aluminum, 20' high	Each	1300
40' high	Each	3125
70' high	Each	9425
Fiberglass, 23' high	Each	1350
39'-5" high	Each	2050
59' high	Each	5150

Description	Unit	$ Cost
LIBRARY FURNISHINGS		
Bookshelf, 90" high, 10" shelf double face	L.F.	118
single face	L.F.	92
Charging desk, built-in with counter		
Plastic laminated top	L.F.	430
Reading table, laminated		
top 60" x 36"	Each	585

Use *Location Factors* in Reference Section

Figure 5–8

More detailed square foot costs for building a library.

Model costs calculated for a 2 story building with 14' story height and 22,000 square feet of floor area.

NO.	SYSTEM/COMPONENT	SPECIFICATIONS		UNIT	UNIT COST	COST PER S.F.	% OF SUB-TOT
1.0 FOUNDATIONS							
.1	Footings & Foundations	Poured concrete; strip and spread footings and 4' foundation wall		S.F. Ground	3.88	1.94	
.4	Piles & Caissons	N/A		–	–	–	3.6%
.9	Excavation & Backfill	Site preparation for slab and trench for foundation wall and footing		S.F. Ground	.92	.46	
2.0 SUBSTRUCTURE							
.1	Slab on Grade	4" reinforced concrete with vapor barrier and granular base		S.F. Slab	2.85	1.42	2.1%
.2	Special Substructures	N/A		–	–	–	
3.0 SUPERSTRUCTURE							
.1	Columns & Beams	Concrete columns		L.F. Column	57	1.75	
.4	Structural Walls	N/A		–	–	–	
.5	Elevated Floors	Concrete waffle slab		S.F. Floor	11.47	5.74	20.0%
.7	Roof	Concrete waffle slab		S.F. Roof	10.72	5.36	
.9	Stairs	Concrete filled metal pan		Flight	4975	.45	
4.0 EXTERIOR CLOSURE							
.1	Walls	Face brick with concrete block backup	90% of wall	S.F. Wall	18.14	9.04	
.5	Exterior Wall Finishes	N/A		–	–	–	
.6	Doors	Double aluminum and glass, single leaf hollow metal		Each	3325	.30	16.5%
.7	Windows & Glazed Walls	Window wall	10% of wall	S.F. Wall	29	1.66	
5.0 ROOFING							
.1	Roof Coverings	Built-up tar and gravel with flashing		S.F. Roof	1.90	.95	
.7	Insulation	Perlite/EPS composite		S.F. Roof	1.04	.52	2.4%
.8	Openings & Specialties	Gravel stop and hatches		L.F. Roof	.22	.11	
6.0 INTERIOR CONSTRUCTION							
.1	Partitions	Gypsum board on metal studs	30 S.F. Floor/L.F. Partition	S.F. Partition	3.80	1.52	
.4	Interior Doors	Single leaf wood	300 S.F. Floor/Door	Each	402	1.34	
.5	Wall Finishes	Paint		S.F. Surface	.46	.37	
.6	Floor Finishes	50% carpet, 50% vinyl tile		S.F. Floor	3.14	3.14	16.4%
.7	Ceiling Finishes	Mineral fiber on concealed zee bars		S.F. Ceiling	3.15	3.15	
.9	Interior Surface/Exterior Wall	Painted gypsum board on furring	90% of wall	S.F. Wall	2.75	1.37	
7.0 CONVEYING							
.1	Elevators	One hydraulic passenger elevator		Each	50,380	2.29	3.4%
.2	Special Conveyors	N/A		–	–	–	
8.0 MECHANICAL							
.1	Plumbing	Toilet and service fixtures, supply and drainage	1 Fixture/1835 S.F. Floor	Each	2458	1.34	
.2	Fire Protection	Wet pipe sprinkler system		S.F. Floor	1.39	1.39	
.3	Heating	Included in 8.4		–	–	–	24.2%
.4	Cooling	Multizone unit, gas heating, electric cooling		S.F. Floor	13.35	13.35	
.5	Special Systems	N/A		–	–	–	
9.0 ELECTRICAL							
.1	Service & Distribution	400 ampere service, panel board and feeders		S.F. Floor	.63	.63	
.2	Lighting & Power	Fluorescent fixtures, receptacles, switches, A.C. and misc. power		S.F. Floor	6.45	6.45	11.4%
.4	Special Electrical	Alarm systems and emergency lighting		S.F. Floor	.48	.48	
11.0 SPECIAL CONSTRUCTION							
.1	Specialties	N/A		–	–	–	0.0%
12.0 SITE WORK							
.1	Earthwork	N/A		–	–	–	
.3	Utilities	N/A		–	–	–	0.0%
.5	Roads & Parking	N/A		–	–	–	
.7	Site Improvements	N/A		–	–	–	
			SUB-TOTAL			66.52	100%
	GENERAL CONDITIONS (Overhead & Profit)				15%	9.98	
	ARCHITECT FEES				8%	6.10	
			TOTAL BUILDING COST			82.60	

BUILDING TYPES

Figure 5–9

Detail breakout of library modeled.

From *Means Square Foot Cost Data 1995*. Copyright R. S. Means Co., Inc., Kingston, MA, 617-585-7880, all rights reserved.

Using the Modeled Square Foot Data

What is the estimated cost of a 15,000 sq.ft. library, using this index? Assume face brick with concrete block back-up and a reinforced concrete frame, again using 1995 data (see Fig. 5–8).

size in sq.ft. × cost per sq.ft. = total cost

Determining the cost per sq.ft. of the library using this guide is a little tricky. The table shows the following estimates for given sizes:

13,000 sq.ft. = $89.00
16,000 sq.ft. = $87.00

For a 15,000 sq.ft. building, interpolate to a cost of $87.67 per square foot. Then to arrive at a total,

15,000 sq.ft. × $87.67 = $1,315,050

Both approaches have provided us with a square foot price that at this point is unadjusted and therefore is good only for the national average city in the data year.

Adjustments

Reported Square Foot Prices

To make adjustments to the reported square foot prices, the same method is used as in a ROM estimate. To design and build a median quality, 15,000 sq.ft. library in Orlando, Florida, for instance, with construction to begin in May 1995, the following adjustments are made. Recall that the cost from above for a median quality job at 1995 prices was determined to be $1,315,500.
First adjust for size (see again Fig. 5–3):

$$\text{Size factor} = \frac{15,000}{12,000} = 1.25$$

Read cost multiplier = .98
Library adjusted for size = $1,315,500 × .98 = $1,289,190

Then we adjust for location:

Orlando index = 88.6
Orlando cost = $1,289,190 × (88.6/100) = $1,142,222

Next the adjustment is made for time. Assuming a June 1995 construction start and a projected increase of 2.5 percent per year, add 1.25 percent to the above (early 1995) price.

$$
\begin{aligned}
\text{Total project cost} \ &= \ \$1{,}142{,}222 + (\$1{,}142{,}222 \times 0.0125) \\
&= \ \$1{,}142{,}222 + \$14{,}278 \\
&= \ \$1{,}156{,}500
\end{aligned}
$$

A further adjustment would add in the design fee (see Fig. 5–10). That reference table shows that a library is in the same category as an apartment building. Therefore read 7.3 percent, as this building's cost is close to $1,000,000.

$$
\begin{aligned}
\text{Total project cost} \ &= \ \$1{,}156{,}500 + (\$1{,}156{,}500 \times .073) \\
&= \ \$1{,}156{,}500 + \$84{,}424 \ \text{design fee} \\
&= \ \$1{,}240{,}924
\end{aligned}
$$

The above represents a median quality 15,000 sq.ft. library in Orlando. Construction would begin in mid 1995, and the cost includes a 7.3 percent fee for design services.

Modeled Square Foot Prices

The price for the same library would be figured somewhat differently using this method because data from Means' Modeled Square Foot reference does not include some common features that would be in a typical library. These have to be added in. Typical add-on features include study carrels, emergency lights, a flagpole, and fur-

General Conditions	**R10.1-100**	**Design & Engineering**

Table 10.1-101   Architectural Fees

Tabulated below are typical percentage fees by project size, for good professional architectural service. Fees may vary from those listed depending upon degree of design difficulty and economic conditions in any particular area.

Rates can be interpolated horizontally and vertically. Various portions of the same project requiring different rates should be adjusted proportionately. For alterations, add 50% to the fee for the first $500,000 of project cost and add 25% to the fee for project cost over $500,000.

Architectural fees tabulated below include Engineering Fees.

Building Types	Total Project Size in Thousands of Dollars						
	100	250	500	1,000	5,000	10,000	50,000
Factories, garages, warehouses, repetitive housing	9.0%	8.0%	7.0%	6.2%	5.3%	4.9%	4.5%
Apartments, banks, schools, libraries, offices, municipal buildings	11.7	10.8	8.5	7.3	6.4	6.0	5.6
Churches, hospitals, homes, laboratories, museums, research	14.0	12.8	11.9	10.9	8.5	7.8	7.2
Memorials, monumental work, decorative furnishings	–	16.0	14.5	13.1	10.0	9.0	8.3

Figure 5–10

Sample design fees, as a percent of project cost.

nishings such as bookshelves, reading tables, and a charging desk. Sitework is not included, but the architect's fee is.

The price determined previously, building with face brick with concrete block backup with a reinforced concrete frame, using 1995 data, was $1,315,050. Accounting for these project additives (see again Fig. 5–8):

(5) Emergency lights @ 610 ea	=	3,050
(20) Study carrels @ 700 ea	=	14,000
(1) Flagpole	=	3,125
Bookshelves (300 lf) @ 118/lf	=	35,400
Charging desk (10 lf) @ 430/lf	=	4,300
(10) Reading tables @ 585 ea	=	5,850
Additives total		$65,725
Total cost		$1,380,775

Note: Remember that the modeled square foot data automatically adjusts for size when the price is taken from the data page. The $87.67 per square foot cost for this 15,000 sq.ft. building is less than the $89.00 per square foot for a 13,000 sq.ft. building.

Now adjust for location:

Orlando cost = $1,380,775 × (88%/100) = $1,223,367

Finally, adjust for time. Assume a midyear 1995 start as above, which would increase the price by 1.25 percent.

$$\text{Total project cost} \; = \; \$1,223,367 + (\$1,223,367 \times 1.25\%)$$
$$= \; \$1,223,367 + \$15,292$$
$$= \; \$1,238,659$$

The above price represents the cost of building a library of median quality with typical additives. The cost adjusts for construction in the middle of 1995 in Orlando and includes a design fee.

Presentation

The square foot estimate, like the ROM, would be developed in the conceptual stage of a project and would be used for budgeting purposes. The library worked in this example includes a design fee which was not included in the ROM example, although it could have been added easily. The adjustments included were the same as those used in the ROM estimate. The major difference is that the square foot estimate works with project square footage versus the number of units, such as motel rooms in the previous ROM example. When using the modeled square foot data the estimator also has the ability to adjust for project specifics such as the frame, skin, perimeter, and roofing material.

This estimate does not include land cost or the cost of demolition. The estimate should take no longer than an hour or two to prepare and provides an accuracy in the range of ±15 percent. The estimator would normally apply an appropriate contingency when presenting the final project price to allow for possible scope changes or unfavorable market issues at the time of construction.

Assemblies Estimating

Approach

Assemblies estimating—also called systems estimating—is best accomplished concurrently with the design phase of a project. The estimate is prepared by working with the system or assembly unit of a project. In a ROM estimate a gross unit is established, for example the number of hospital beds required. In square foot estimating the estimator works with the project area. In assemblies estimating the estimator will use more detailed units such as square feet of partition wall, numbers of plumbing fixtures, or square feet of carpet. Since the units are smaller, the estimate becomes more flexible and accurate, but it requires greater designer input and therefore takes longer to prepare. The first system estimates will be accomplished during the schematic design stage and will generally take a day or longer, providing an accuracy in the ±10 percent range.

As a project design develops, the systems estimate needs to be updated to inform the project team of design decision impacts. In an assemblies estimate quality is now treated by the specific material or method chosen, and no longer by using ranges such as ¼ or ¾. Provided good data is available, the designer should be able to look at the costs of different alternatives and make a selection based on cost as well as durability and owner value. Assemblies estimating is an essential part of a value engineering program. As an example, consider the decision as to what type of floor covering to use in an office building:

Floor area = 10,000 sq.ft.

The three best alternatives seem to be:

			Total
Woven wool carpet (42 oz) with padding	=	$6.68/sq.ft.	$66,680
Nylon carpet (26 oz)	=	$3.06/sq.ft.	$30,600
Resilient asphalt tile	=	$1.57/sq.ft.	$15,700

As can be seen, the project cost savings of choosing resilient flooring over the high quality carpet is over $50,000. However, initial cost is only one factor to consider. There are also ongoing maintenance issues to look at, along with aesthetics, acoustics, and the intended usage. This process of working through the design of a project and making design decisions cognizant of the cost implications of these decisions is the real value of systems estimating.

The systems estimate also provides a yardstick for the comparison of the costs of different subsystems within a project. As an example, a commercial building project is estimated by following what is called the uniformat breakdown (see Fig. 5–11). By comparing the twelve divisional calculated costs with the average for past projects, the project team can verify the estimate's accuracy both overall and by division, as well as get a sense of the scope of the project. If a division (as a percent of the project total) is much higher than normal, the project team should investigate why. The table in Figure 5–12 provides a typical twelve-division breakdown for a general commercial building.

Many projects are large and are designed by large design teams. The process of using the systems estimate to check and verify the design to date should serve to keep the project balanced and within budget and responsive to the owner's needs.

Data

The information needed for a systems estimate must be available in a form that can be quickly itemized on a system-by-system basis. Using an interior partition as an example, the partition, which will be estimated per square foot, must include the

Figure 5–11
Specification formats.

Masterformat	Uniformat
1. GEN. REQUIREMENTS	1. FOUNDATION
2. SITE WORK	2. SUBSTRUCTURE
3. CONCRETE	3. SUPERSTRUCTURE
4. MASONRY	4. EXTERIOR CLOSURE
5. METALS	5. ROOFING
6. WOOD & PLASTICS	6. INTERIOR CONSTR.
7. MOISTURE PROTECTION	7. CONVEYING
8. DOORS, WINDOWS & GLASS	8. MECHANICAL
9. FINISHES	9. ELECTRICAL
10. SPECIALTIES	10. GEN. CONDITIONS
11. ARCHIT. EQUIPMENT	11. SPECIALTIES
12. FURNISHINGS	12. SITE WORK
13. SPECIAL CONSTRUCTION	
14. CONVEYING SYSTEMS	
15. MECHANICAL	
16. ELECTRICAL	

Table 13.1-011 Labor, Material and Equipment Cost Distribution for Weighted Average Listed by System Division

Division No.	Building System	Percentage	Division No.	Building System	Percentage
1 & 2	Foundation & Substructure	6.5 %	7	Conveying	3.4%
3	Superstructure	17.8	8	Mechanical	21.4
4	Exterior Closure	12.1	9	Electric	12.0
5	Roofing	2.7	11	Equipment	1.8
6	Interior Construction	17.2	12	Site Work	5.1
				Total weighted index (Div. 1-12)	100.0%

Figure 5–12

Typical cost distributions by division.

cost of the metal studs, drywall installation, and the taping and finishing. It would be useful to have in the company data base partition options including different fire ratings, insulated vs. noninsulated, and partitions finished both sides vs. only one side. Estimating software packages (Timberline, for example) allow the estimator to create different assemblies.

The R.S. Means company publishes an Assemblies Cost Data book on an annual basis which will be used for example purposes in this section. The book is organized in accordance with the twelve uniformat divisions illustrated earlier. Referencing Figures 5–13 and 5–14, notice the information illustrated for an interior drywall partition system. The estimator must (1) identify the partition system that will be used, and (2) determine the number of square feet of partition that will be needed. As can be seen in the illustration, the estimator in conjunction with the designer must decide between wood vs. metal studs, and fire resistant vs. water resistant drywall on the face layer, and decide on the base layer drywall type, stud spacing, and whether or not insulation will be used. The corresponding cost for installation and material for each option can be read to the right. These costs are the installation costs including overhead and profit for the installing contractor.

For example purposes, assume that the drywall specified in line #6.1-510-1450 from the table in Figure 5–13 was selected for use in a 10,000 sq.ft. office area.

At this point the estimator, if drawings are available, can measure the partition square footage (linear feet × height) or by using an approximation or estimating aid such as Figure 5–15 arrive at the quantity of drywall necessary for the project.

Assuming a two-story 10,000 square foot building, and using the numbers from Figure 5–15:

10,000 sq.ft./20 sq.ft. per lf = 500 lf of partition

70% drywall = 350 lf

30% block = 150 lf

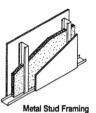

The Drywall Partitions/Stud Framing Systems are defined by type of drywall and number of layers, type and spacing of stud framing, and treatment on the opposite face. Components include taping and finishing.

Cost differences between regular and fire resistant drywall are negligible, and terminology is interchangeable. In some cases fiberglass insulation is included for additional sound deadening.

Wood Stud Framing   **Metal Stud Framing**

System Components		QUANTITY	UNIT	COST PER S.F.		
				MAT.	INST.	TOTAL
SYSTEM 6.1-510-1250 **DRYWALL PARTITION,5/8" F.R.1 SIDE,5/8" REG.1 SIDE,2"X4"STUDS,16" O.C.**						
Gypsum plasterboard, nailed/screwed to studs, 5/8"F.R. fire resistant		1.000	S.F.	.24	.31	.55
Gypsum plasterboard, nailed/screwed to studs, 5/8" regular		1.000	S.F.	.22	.31	.53
Taping and finishing joints		2.000	S.F.	.16	.62	.78
Framing, 2 x 4 studs @ 16" O.C., 10' high		1.000	S.F.	.45	.65	1.10
	TOTAL			1.07	1.89	2.96

6.1-510 — Drywall Partitions/Wood Stud Framing

	FACE LAYER	BASE LAYER	FRAMING	OPPOSITE FACE	INSULATION	COST PER S.F.		
						MAT.	INST.	TOTAL
1200	5/8" FR drywall	none	2 x 4, @ 16" O.C.	same	0	1.09	1.89	2.98
1250				5/8" reg. drywall	0	1.07	1.89	2.96
1300				nothing	0	.77	1.27	2.04
1400		1/4" SD gypsum	2 x 4 @ 16" O.C.	same	1-1/2" fiberglass	1.77	2.90	4.67
1450				5/8" FR drywall	1-1/2" fiberglass	1.59	2.55	4.14
1500				nothing	1-1/2" fiberglass	1.27	1.93	3.20
1600		resil. channels	2 x 4 @ 16", O.C.	same	1-1/2" fiberglass	1.63	3.68	5.31
1650				5/8" FR drywall	1-1/2" fiberglass	1.52	2.94	4.46
1700				nothing	1-1/2" fiberglass	1.20	2.32	3.52
1800		5/8" FR drywall	2 x 4 @ 24" O.C.	same	0	1.47	2.38	3.85
1850				5/8" FR drywall	0	1.23	2.07	3.30
1900				nothing	0	.91	1.45	2.36
1950		5/8" FR drywall	2 x 4, 16" O.C.	same	0	1.57	2.51	4.08
1955				5/8" FR drywall	0	1.33	2.20	3.53
2000				nothing	0	1.01	1.58	2.59
2010		5/8" FR drywall	staggered, 6" plate	same	0	2.04	3.17	5.21
2015				5/8" FR drywall	0	1.80	2.86	4.66
2020				nothing	0	1.48	2.24	3.72
2200		5/8" FR drywall	2 rows-2 x 4	same	2" fiberglass	2.40	3.47	5.87
2250			16"O.C.	5/8" FR drywall	2" fiberglass	2.16	3.16	5.32
2300				nothing	2" fiberglass	1.84	2.54	4.38
2400	5/8" WR drywall	none	2 x 4, @ 16" O.C.	same	0	1.25	1.89	3.14
2450				5/8" FR drywall	0	1.17	1.89	3.06
2500				nothing	0	.85	1.27	2.12
2600		5/8" FR drywall	2 x 4, @ 24" O.C.	same	0	1.63	2.38	4.01
2650				5/8" FR drywall	0	1.31	2.07	3.38
2700				nothing	0	.99	1.45	2.44
2800	5/8 VF drywall	none	2 x 4, @ 16" O.C.	same	0	1.75	2.05	3.80
2850				5/8" FR drywall	0	1.42	1.97	3.39
2900				nothing	0	1.10	1.35	2.45

Figure 5–13

Assemblies cost data for drywall partitions.

From *Means Assemblies Cost Data 1995*. Copyright R. S. Means Co., Inc., Kingston, MA, 617-585-7880, all rights reserved.

6.1-510 — Drywall Partitions/Wood Stud Framing

	FACE LAYER	BASE LAYER	FRAMING	OPPOSITE FACE	INSULATION	COST PER S.F.		
						MAT.	INST.	TOTAL
3000	5/8 VF drywall	5/8" FR drywall	2 x 4 , 24" O.C.	same	0	2.13	2.54	4.67
3050				5/8" FR drywall	0	1.56	2.15	3.71
3100				nothing	0	1.24	1.53	2.77
3200	1/2" reg drywall	3/8" reg drywall	2 x 4, @ 16" O.C.	same	0	1.37	2.51	3.88
3250				5/8" FR drywall	0	1.23	2.20	3.43
3300				nothing	0	.91	1.58	2.49

6.1-510 — Drywall Partitions/Metal Stud Framing

	FACE LAYER	BASE LAYER	FRAMING	OPPOSITE FACE	INSULATION	COST PER S.F.		
						MAT.	INST.	TOTAL
5200	5/8" FR drywall	none	1-5/8" @ 24" O.C.	same	0	.85	1.84	2.69
5250				5/8" reg. drywall	0	.83	1.84	2.67
5300				nothing	0	.53	1.22	1.75
5400			3-5/8" @ 24" O.C.	same	0	.93	1.87	2.80
5450				5/8" reg. drywall	0	.91	1.87	2.78
5500				nothing	0	.61	1.25	1.86
5600		1/4" SD gypsum	1-5/8" @ 24" O.C.	same	0	1.21	2.54	3.75
5650				5/8" FR drywall	0	1.03	2.19	3.22
5700				nothing	0	.71	1.57	2.28
5800			2-1/2" @ 24" O.C.	same	0	1.24	2.56	3.80
5850				5/8" FR drywall	0	1.06	2.21	3.27
5900				nothing	0	.74	1.59	2.33
6000		5/8" FR drywall	2-1/2" @ 16" O.C.	same	0	1.64	2.57	4.21
6050				5/8" FR drywall	0	1.40	2.26	3.66
6100				nothing	0	1.08	1.64	2.72
6200			3-5/8" @ 24" O.C.	same	0	1.41	2.49	3.90
6250				5/8"FR drywall	3-1/2" fiberglass	1.41	2.38	3.79
6300				nothing	0	.85	1.56	2.41
6400	5/8" WR drywall	none	1-5/8" @ 24" O.C.	same	0	1.01	1.84	2.85
6450				5/8" FR drywall	0	.93	1.84	2.77
6500				nothing	0	.61	1.22	1.83
6600			3-5/8" @ 24" O.C.	same	0	1.09	1.87	2.96
6650				5/8" FR drywall	0	1.01	1.87	2.88
6700				nothing	0	.69	1.25	1.94
6800		5/8" FR drywall	2-1/2" @ 16" O.C.	same	0	1.80	2.57	4.37
6850				5/8" FR drywall	0	1.48	2.26	3.74
6900				nothing	0	1.16	1.64	2.80
7000			3-5/8" @ 24" O.C.	same	0	1.57	2.49	4.06
7050				5/8"FR drywall	3-1/2" fiberglass	1.49	2.38	3.87
7100				nothing	0	.93	1.56	2.49
7200	5/8" VF drywall	none	1-5/8" @ 24" O.C.	same	0	1.51	2	3.51
7250				5/8" FR drywall	0	1.18	1.92	3.10
7300				nothing	0	.86	1.30	2.16
7400			3-5/8" @ 24" O.C.	same	0	1.59	2.03	3.62
7450				5/8" FR drywall	0	1.26	1.95	3.21
7500				nothing	0	.94	1.33	2.27
7600		5/8" FR drywall	2-1/2" @ 16" O.C.	same	0	2.30	2.73	5.03
7650				5/8" FR drywall	0	1.73	2.34	4.07
7700				nothing	0	1.41	1.72	3.13
7800			3-5/8" @ 24" O.C.	same	0	2.07	2.65	4.72
7850				5/8"FR drywall	3-1/2" fiberglass	1.74	2.46	4.20
7900				nothing	0	1.18	1.64	2.82

Figure 5–14

More assemblies cost data for drywall.

INTERIOR CONSTRUCTION   **6**

Table 15.2-201 Partition/Door Density

Building Type		Stories	Partition/Density	Doors	Description of Partition
Apartments		1 story	9 SF/LF	90 SF/door	Plaster, wood doors & trim
		2 story	8 SF/LF	80 SF/door	Drywall, wood studs, wood doors & trim
		3 story	9 SF/LF	90 SF/door	Plaster, wood studs, wood doors & trim
		5 story	9 SF/LF	90 SF/door	Plaster, wood studs, wood doors & trim
		6-15 story	8 SF/LF	80 SF/door	Drywall, wood studs, wood doors & trim
Bakery		1 story	50 SF/LF	500 SF/door	Conc. block, paint, door & drywall, wood studs
		2 story	50 SF/LF	500 SF/door	Conc. block, paint, door & drywall, wood studs
Bank		1 story	20 SF/LF	200 SF/door	Plaster, wood studs, wood doors & trim
		2-4 story	15 SF/LF	150 SF/door	Plaster, wood studs, wood doors & trim
Bottling Plant		1 story	50 SF/LF	500 SF/door	Conc. block, drywall, wood studs, wood trim
Bowling Alley		1 story	50 SF/LF	500 SF/door	Conc. block, wood & metal doors, wood trim
Bus Terminal		1 story	15 SF/LF	150 SF/door	Conc. block, ceramic tile, wood trim
Cannery		1 story	100 SF/LF	1000 SF/door	Drywall on metal studs
Car Wash		1 story	18 SF/LF	180 SF/door	Concrete block, painted & hollow metal door
Dairy Plant		1 story	30 SF/LF	300 SF/door	Concrete block, glazed tile, insulated cooler doors
Department Store		1 story	60 SF/LF	600 SF/door	Drywall, wood studs, wood doors & trim
		2-5 story	60 SF/LF	600 SF/door	30% concrete block, 70% drywall, wood studs
Dormitory		2 story	9 SF/LF	90 SF/door	Plaster, concrete block, wood doors & trim
		3-5 story	9 SF/LF	90 SF/door	Plaster, concrete block, wood doors & trim
		6-15 story	9 SF/LF	90 SF/door	Plaster, concrete block, wood doors & trim
Funeral Home		1 story	15 SF/LF	150 SF/door	Plaster on concrete block & wood studs, paneling
		2 story	14 SF/LF	140 SF/door	Plaster, wood studs, paneling & wood doors
Garage Sales & Service		1 story	30 SF/LF	300 SF/door	50% conc. block, 50% drywall, wood studs
Hotel		3-8 story	9 SF/LF	90 SF/door	Plaster, conc. block, wood doors & trim
		9-15 story	9 SF/LF	90 SF/door	Plaster, conc. block, wood doors & trim
Laundromat		1 story	25 SF/LF	250 SF/door	Drywall, wood studs, wood doors & trim
Medical Clinic		1 story	6 SF/LF	60 SF/door	Drywall, wood studs, wood doors & trim
		2-4 story	6 SF/LF	60 SF/door	Drywall, wood studs, wood doors & trim
Motel		1 story	7 SF/LF	70 SF/door	Drywall, wood studs, wood doors & trim
		2-3 story	7 SF/LF	70 SF/door	Concrete block, drywall on wood studs, wood paneling
Movie Theater	200-600 seats	1 story	18 SF/LF	180 SF/door	Concrete block, wood, metal, vinyl trim
	601-1400 seats		20 SF/LF	200 SF/door	Concrete block, wood, metal, vinyl trim
	1401-22000 seats		25 SF/LF	250 SF/door	Concrete block, wood, metal, vinyl trim
Nursing Home		1 story	8 SF/LF	80 SF/door	Drywall, wood studs, wood doors & trim
		2-4 story	8 SF/LF	80 SF/door	Drywall, wood studs, wood doors & trim
Office		1 story	20 SF/LF	200-500 SF/door	30% concrete block, 70% drywall on wood studs
		2 story	20 SF/LF	200-500 SF/door	30% concrete block, 70% drywall on wood studs
		3-5 story	20 SF/LF	200-500 SF/door	30% concrete block, 70% movable partitions
		6-10 story	20 SF/LF	200-500 SF/door	30% concrete block, 70% movable partitions
		11-20 story	20 SF/LF	200-500 SF/door	30% concrete block, 70% movable partitions
Parking Ramp (Open)		2-8 story	60 SF/LF	600 SF/door	Stair and elevator enclosures only
Parking garage		2-8 story	60 SF/LF	600 SF/door	Stair and elevator enclosures only
Pre-Engineered	Steel	1 story	0		
	Store	1 story	60 SF/LF	600 SF/door	Drywall on wood studs, wood doors & trim
	Office	1 story	15 SF/LF	150 SF/door	Concrete block, movable wood partitions
	Shop	1 story	15 SF/LF	150 SF/door	Movable wood partitions
	Warehouse	1 story	0		
Radio & TV Broadcasting		1 story	25 SF/LF	250 SF/door	Concrete block, metal and wood doors
& TV Transmitter		1 story	40 SF/LF	400 SF/door	Concrete block, metal and wood doors
Self Service Restaurant		1 story	15 SF/LF	150 SF/door	Concrete block, wood and aluminum trim
Cafe & Drive-in Restaurant		1 story	18 SF/LF	180 SF/door	Drywall, wood studs, ceramic & plastic trim
Restaurant with seating		1 story	25 SF/LF	250 SF/door	Concrete block, paneling, wood studs & trim
Supper Club		1 story	25 SF/LF	250 SF/door	Concrete block, paneling, wood studs & trim
Bar or Lounge		1 story	24 SF/LF	240 SF/door	Plaster or gypsum lath, wooded studs
Retail Store or Shop		1 story	60 SF/LF	600 SF/door	Drywall wood studs, wood doors & trim
Service Station	Masonry	1 story	15 SF/LF	150 SF/door	Concrete block, paint, door & drywall, wood studs
	Metal panel	1 story	15 SF/LF	150 SF/door	Concrete block, paint, door & drywall, wood studs
	Frame	1 story	15 SF/LF	150 SF/door	Drywall, wood studs, wood doors & trim
Shopping Center	(strip)	1 story	30 SF/LF	300 SF/door	Drywall, wood studs, wood doors & trim
	(group)	1 story	40 SF/LF	400 SF/door	50% concrete block, 50% drywall, wood studs
		2 story	40 SF/LF	400 SF/door	50% concrete block, 50% drywall, wood studs
Small Food Store		1 story	30 SF/LF	300 SF/door	Concrete block drywall, wood studs, wood trim
Store/Apt. above	Masonry	2 story	10 SF/LF	100 SF/door	Plaster, wood studs, wood doors & trim
	Frame	2 story	10 SF/LF	100 SF/door	Plaster, wood studs, wood doors & trim
	Frame	3 story	10 SF/LF	100 SF/door	Plaster, wood studs, wood doors & trim
Supermarkets		1 story	40 SF/LF	400 SF/door	Concrete block, paint, drywall & porcelain panel
Truck Terminal		1 story	0		
Warehouse		1 story	0		

Figure 5–15

Reference aids for drywall assemblies.

REFERENCE AIDS

Assume a 12' height:

sq.ft. drywall = 350 lf × 12' = 4,200 sq.ft.

sq.ft. block = 150 lf × 12' = 1,800 sq.ft.

Total drywall cost = Quantity drywall in sq.ft. × Cost per sq.ft.

Total drywall cost = 4,200 sq.ft. × $4.14 per sq.ft. = $17,388

Existing drawings and/or estimating aids and approximations would be used to determine the quantities for all of the project components. Assembly cost data would be taken from company historical records, from data books such as Means, from suppliers and vendors, or would be created by the estimator by the use of unit prices. Once all of the quantities and system unit prices are determined, the total assembly estimate can be produced.

Compilation and Adjustment

Each of the major elements of the project would be quantified and priced by its major assemblies as the interior partitions were just treated. For a commercial building project the assemblies would be summarized as illustrated in the following list (see Fig. 5–16). The major elements for each of these divisions are as follows:

Div 1	Foundations and excavation
Div 2	Slab on grade
Div 3	Floor and roof structures and superstructure
Div 4	Building envelope—windows, doors, and walls
Div 5	Roofing—membrane, insulation, and flashing
Div 6	Partitions, interior doors, finish floors, and ceilings
Div 7	Elevators, escalators, and dumbwaiters
Div 8	Plumbing, heating and cooling, and fire protection
Div 9	Service, power, and lighting
Div 10	General conditions
Div 11	Architectural equipment and furnishings
Div 12	Excavation, roadways and parking, and landscaping

For projects other than commercial buildings the divisions would be organized differently, in accordance to the method in which the project is structured. A highway project for instance, might be organized along major elements such as clearing and grading, paving, drainage, and bridge abutments. Depending on the information available, general conditions, Division 10 above, could be itemized, but would more likely be treated as a percentage of construction cost. Sales tax, where applicable, would be figured as a percentage of material costs, which is approximately 50 percent of total project cost. Home office overhead is usually treated as a percentage,

▲ Means Forms
PRELIMINARY
ESTIMATE (Cost Summary)

PROJECT		TOTAL AREA	SHEET NO.
LOCATION		TOTAL VOLUME	ESTIMATE NO
ARCHITECT		COST PER S.F.	DATE
OWNER		COST PER C.F.	NO. OF STORIES
QUANTITIES BY:	PRICES BY:	EXTENSIONS BY:	CHECKED BY:

NO.	DESCRIPTION	SUB TOTAL COST	COST/S.F.	%
1.0	Foundation			
2.0	Substructure			
3.0	Superstructure			
4.0	Exterior Closure			
5.0	Roofing			
6.0	Interior Construction			
7.0	Conveying			
8.0	Mechanical System			
9.0	Electrical			
10.0	General Conditions (Breakdown)			
11.0	Special Construction			
12.0	Site Work			

Building Sub Total $ _____

Sales Tax _____ % × Sub Total $ _____ /2 =

$ _____

$ _____

General Conditions (%) _____ % × Sub Total $ _____ =

General Conditions $ _____

Sub Total "A" $ _____

Overhead_____ % × Sub Total "A" $ _____ =

Sub Total "B" $ _____

Profit_____ % × Sub Total "B" $ _____ =

Sub Total "C" $ _____

Location Factor _____ % × Sub Total "C" $ _____ =

Adjusted Building Cost $ _____

Architects Fee_____ % × Adjusted Building Cost $ _____ = $ _____

Contingency_____ % × Adjusted Building Cost $ _____ = $ _____

Total Cost [_____]

Square Foot Cost $ _____ / _____ S.F. = _____ $/S.F.
Cubic Foot Cost $ _____ / _____ C.F. = _____ $/C.F.

Figure 5–16
Systems costs summary form.
From *Means Assemblies Cost Data 1995*. Copyright R. S. Means Co., Inc., Kingston, MA, 617-585-7880, all rights reserved.

as is profit. The percentage used for home office overhead and profit is dependent on the amount of work the company has in place, the efficiency of the company, the perceived difficulty of the project, and the amount of competition expected, as well as many other market conditions and factors.

As is shown on the worksheet in Figure 5–16, the project should be adjusted for location as well as time. If the estimate needs to reflect total project cost to the owner, the designer's fee needs to be added in as well. Project contingency would be added in at the end to allow some room for adjustment in owner scope.

Presentation

An assemblies estimate might be done several times throughout the course of the design and should be an integral part of the design process. A formal review of the estimate should occur at the end of schematic design and again at the end of the design development phase. The schematic estimate would be a complete assemblies estimate, while the design development estimate would include some assemblies pricing combined with some unit pricing (which is the subject of the next chapter).

It is important in the estimate presentation to highlight the design elements that are generating the greatest project cost, particularly work items that have some degree of flexibility as to owner choice. Required code items would have little flexibility, whereas fine millwork, high end carpet, or high speed elevator systems may have alternate options. Identifying these elements early provides the opportunity for cost savings or an early adjustment in the project's focus that may end up providing the owner greater value at less cost.

The key, however, is to get an accurate, early budget for the project, verifying that the design is within or that it exceeds the owner's budget. If the design exceeds the owner's budget, design alternatives may be investigated, the scope of the project reduced, or additional funding pursued. The earlier and more accurate the estimate, the greater the opportunity for such decisions to be made.

Conclusion

This chapter has reviewed the three types of estimates that would be conducted in the early stages of a project. The conceptual/ROM estimate would be conducted early, be used primarily for budgetary reasons, be done quickly, and provide an accuracy in the ±20 percent range. The square foot estimate would also be done primarily in the conceptual stage, would take slightly longer than a ROM estimate, and would provide an accuracy in the ±15 percent range. The assemblies estimate would be done concurrently with the project's design. It would utilize design information, take longer to prepare, and provide an accuracy in the ±10 percent range.

The ROM and square foot estimates can both be done with little if any design information, whereas the assemblies estimate needs at a minimum design detail approaching the 25–30 percent complete range. As the number of project decisions increase, the amount of detail that needs to be quantified and priced increases as well, providing better accuracy, but taking longer to accomplish. This chapter used information provided primarily by the R.S. Means company, but most larger owners, designers, and construction companies, through historical records, develop their own assembly costs which they use for future estimates.

The process that was followed in conducting all of the estimates discussed was similar in that each estimate considered the five major elements of size, quality, location, time, and market. In the ROM estimate the size was considered by units, quality by ¼ vs. ¾, location and time were treated by indices, and market would be figured into the contingency. The square foot estimate considered project area, while the remaining items were all treated the same as in the ROM. In the assemblies estimate the size of the project was quantified by assemblies, and quality was considered by selecting the specific system desired; location and time were factored by indices, and other market considerations would be factored into overhead and profit percentages. The unit price estimate, the last estimate type to be covered, will be the subject of the next chapter.

Chapter Review Questions

1. A rough order of magnitude estimate would be used for bidding purposes.
   ___ T   ___ F

2. An index measures change with respect to an established baseline.
   ___ T   ___ F

3. When adjusting for time and location, it is necessary that the adjustment for location be made first.
   ___ T   ___ F

4. Quality is considered in a reported square foot estimate by using the ¼, median, and ¾ costs.
   ___ T   ___ F

5. Systems or assemblies estimating is an important tool in a value engineering program.
   ___ T   ___ F

6. The purpose of the cost multiplier used in the size adjustment step is _____.
   a. To adjust for market conditions.
   b. To adjust for economy of scale.
   c. To allow for the increase in cost per square foot as projects increase in size.
   d. All of the above.

7. City cost indices allow an adjustment for _____.
   a. Regional changes in the cost of labor.
   b. Regional changes in materials.
   c. Regional changes in equipment.
   d. All of the above.

8. The modeled square foot estimating approach is based on _____.
   a. Actual costs of past projects.
   b. Calculations of the estimator who has figured actual quantities and multiplied these by reported square foot prices.
   c. Typical building types that are computer modeled and adjusted annually for price.
   d. None of the above.

9. The systems or assemblies estimate adjusts for quality by _____.
   a. The specific method or material chosen.
   b. The use of ¼, median, and ¾ prices.
   c. The use of quality indices.
   d. All of the above.

10. Which of the following properly summarizes the estimate types in increased order of accuracy?
   a. Square foot, assemblies, and ROM
   b. ROM, assemblies, and square foot
   c. Assemblies, square foot, and ROM
   d. ROM, square foot, and assemblies

Exercises

1. What is the projected return on investment (ROI) for a 200-room motel to be built in Norfolk, Virginia? Use Means reported square foot data. Consider the following assumptions:
   a. Median level of quality
   b. 85 percent occupancy
   c. Nightly rate of $85
   d. 10 percent profit
   e. Spring 1995 construction

   $$\text{Return on investment} = \frac{\text{Dollar amount earned}}{\text{Dollar amount invested}}$$

2. You have been asked to decide between two roofing membrane options for a new project. Option 1 is to use a 4-ply #15 asphalt felt, mopped. Option 2 calls for PVC, 48 mils, fully adhered with adhesive. Option 1 is less expensive, but you estimate it will cost 10¢/square foot in additional maintenance over the life of the building. Both are projected to last 15 years. Which is the better choice?

CHAPTER

6

Detailed Estimating

STUDENT LEARNING OBJECTIVES

From studying this chapter, you will learn:

1. Why unit price estimates are prepared, and how they are used
2. How to accomplish a quantity takeoff
3. How to obtain unit prices
4. How to set up a detailed estimate and figure overhead and profit

Introduction

The detailed estimate, also called a unit price estimate, is the last type of estimate that will be discussed in this book. Detailed estimates are typically prepared towards the end of the design phase, as they require precise project information. A detailed estimate which is prepared by the project team will be used to determine the fair cost for the project and is hence called a **fair cost estimate.** Contractors who are bidding the project also prepare detailed estimates. These estimates are called **bid estimates** and, when accepted by the owner, form the contract price for the project.

Detailed estimates take weeks to prepare and involve many people from many different disciplines. A general contractor who is preparing a bid will request proposals from subcontractors for the work that is being subcontracted out, and will quantify ("take off") and price work that will be done by its own work forces. Quotes from material suppliers, also called vendors, will be utilized where possible to get precise material prices. Where quotes cannot be obtained, contractors will utilize company records and published cost data. The contractor must obtain good information on projected wage rates and needs to precisely figure worker productivity to calculate durations.

Good organization is a key to preparing reliable estimates and avoiding mistakes. Estimates are generally broken down by bid packages, so the subcontractor bids received must be tied to the appropriate package. Each subcontractor bid received must be checked to ensure that the bid is accurate and includes the precise work requested—no more, no less. Work that is to be done by the contractor's own work forces, also termed **"in-house"** work, must be properly quantified and checked. Material quotes received must also be verified. How long is the quote good for? Does it include all the items requested? Labor unit prices must be calculated and projected to the time frame when the work will be accomplished.

The total of all the subcontracted work and in-house work, materials, labor, and equipment is called the direct cost of the project. To this total must be added sales tax, the cost of bonding, if required by the owner, and the costs required to manage the job in the field and at the home office. These costs are often called indirect costs; they were described in the sidebar in Chapter 4. The contractor must also add in expected profit.

This chapter will look first at how to determine the quantities involved on a project, how to organize the takeoff process, and how to measure and break down the data so proper unit prices can be attached. Next, the steps involved in pricing will be examined, as will where pricing data and wage rates can be obtained, and how productivity can be determined. Lastly, the proper formatting of an estimate will be discussed. The steps involved in calculating general conditions, as well as overhead and profit, and how to apply these to the estimate will be covered at that time.

Quantity Takeoff

Organization

Preparing an estimate, particularly on a large project, is a complex organizational task. Many people will be involved, as well as a variety of documents. Information will be coming from a number of different sources and will have to be catalogued in a variety of ways. To put together a competitive price while minimizing the number of company hours is the challenge of the estimating team. The key to accomplishing this goal is a good organizational structure. Initially, the team will ensure that adequate space, assignments, and forms and procedures are developed and maintained. A notebook, logbook, or file system will be set up to track all activities that will occur throughout the estimating process. Computerized estimating software is available to accomplish this.

The foundation of any detailed estimate is the contract documents. Particular care will be given to these by the estimating team. It is important to maintain an adequate number of sets so that interested subcontractors and material suppliers can review the project. Drawing and specification sets are generally available from the architect or engineer for a refundable deposit. Most contractors create a plan room either at their office or some central location where interested subcontractors and vendors can review the documents. When the drawings and specifications are received, the first task is to verify that a complete set has been received. The drawings are numbered by discipline, and the specifications separated by trade. All the addenda, which are drawings and specifications issued during the bidding process, also need to be tracked and verified.

It is important for the estimator to understand the project; what it will look like, how it will be constructed, how the trades will move through the project, and what the work environment will be. The drawings will provide a good understanding of the magnitude and scale of the project, while a site analysis will provide a good overview of the work environment. Estimators study the drawings to visualize how the project will be constructed. They look at materials used, the amount of repetition in the project, structural systems used, and mechanical and electrical systems required.

Analysis of the site is equally important. Information about conditions around the construction area will not generally be shown on the drawings. Gathering of this information will require visits to the site. Figure 6–1, a site analysis form, provides a checklist

JOB SITE ANALYSIS

		SHEET NO.
PROJECT		BID DATE
LOCATION		NEAREST TOWN
ARCHITECT	ENGINEER	OWNER

Access, Highway	Surface	Capacity
Railroad Siding	Freight Station	Bus Station
Airport	Motels/Hotels	Hospital
Post Office	Communications	Police
Distance & Travel Time to Site		Dock Facilities
Water Source	Amount Available	Quality
Distance from Site	Pipe/Pump Required?	Tanks Required?
Owner	Price (MG)	Treatment Necessary?
Natural Water Availability		Amount
Power Availability	Location	Transformer
Distance	Amount Available	
Voltage Phase	Cycle	KWH or HP Rate

Temporary Roads	Lengths & Widths	
Bridges/Culverts	Number & Size	
Drainage Problems		
Clearing Problems		
Grading Problems		
Fill Availability	Distance	
Mobilization Time	Cost	
Camps or Housing	Size of Work Force	
Sewage Treatment		
Material Storage Area	Office & Shed Area	

Labor Source	Union Affiliation
Common Labor Supply	Skilled Labor Supply
Local Wage Rates	Fringe Benefits
Travel Time	Per Diem

Taxes, Sales	Facilities	Equipment
Hauling	Transportation	Property
Other		

Material Availability: Aggregates	Cement
Ready Mix Concrete	
Reinforcing Steel	Structural Steel
Brick & Block	Lumber & Plywood
Building Supplies	Equipment Repair & Parts

Demolition: Type	Number	
Size	Equip. Required	
Dump Site	Distance	Dump fees
Permits		

Figure 6–1

Site analysis form

From *Means Form Book.* Copyright R. S. Means Co., Inc., Kingston, MA, 617-585-7880, all rights reserved.

130

Clearing: Area _____ | Timber _____ | Diameter _____ | Species _____

Brush Area _____ | Burn on Site _____ | Disposal Area _____

Saleable Timber _____ | Useable Timber _____ | Haul _____

Equipment Required _____

Weather: Mean Temperatures _____

Highs _____ | Lows _____

Working Season Duration _____ | Bad Weather Allowance _____

Winter Construction _____

Average Rainfall _____ | Wet Season _____ | Dry Season _____

Stream or Tide Conditions _____

Haul Road Problems _____

Long Range Weather _____

Soils: Job Borings Adequate? _____ | Test Pits _____

Additional Borings Needed _____ | Location _____ | Extent _____

Visible Rock _____

U.S. Soil & Agriculture Maps _____

Bureau of Mines Geological Data _____

County/State Agriculture Agent _____

Tests Required _____

Ground Water _____

Construction Plant Required _____

Alternate Method _____

Equipment Available _____

Rental Equipment _____ | Location _____

Miscellaneous: Contractor Interest _____

Sub Contractor Interest _____

Material Fabricator Availability _____

Possible Job Delays _____

Political Situation _____

Construction Money Availability _____

Unusual Conditions _____

Summary _____

_____

_____

_____

_____

_____

_____

_____

_____

_____

_____

_____

131

of the kinds of issues that should be understood about the site before beginning an estimate. The composition of the soil, for example, will affect how easily it can be worked. An estimator needs to look at how the site is accessed and locate the nearest power, water, and phone lines. The area around the site is important. If it is congested, money and time will be spent in phasing deliveries and minimizing storage. Building sites that are close to occupied areas might also mean money for **mitigation,** that is, money spent to lessen the severity of the construction on surrounding people and businesses. Signs, temporary facilities or roadways, and fencing or barriers can all help soften the impact of the project for the public. In renovation projects these issues are even more critical (see sidebar, Repair and Remodeling Projects). On heavy engineering projects truck access and egress can be a major issue affecting when and how much material can be moved to and from the site. On the Central Artery project in Boston, a $7+ billion transportation project, barges and trucks were used to remove excavation material from the site. Truck traffic, however, was prohibited from using the city streets, requiring the construction of a new haul road to accommodate the trucks (see Fig. 6–2).

If the area is unfamiliar, the estimator has to look beyond the immediate site and investigate other issues, such as the availability of local materials and labor, and test the subcontractor interest. A determination must be made as to what work will be subcontracted out and what work will be done in-house. For the work being bid out, clear definable work packages must be prepared. Any significant site services provided by the general contractor or construction manager need to be identified as well, so that double pricing is avoided.

Other issues addressed at this time are bonding and insurance required by the owner, duration of the project, and field conditions and administration requirements. All of these can have significant effect on the bid.

Figure 6–2
A "haul road" needed to remove truck traffic from local city streets. This roadway had to be built prior to construction.
Photo by Don Farrell

Sidebar ▬▬▬▬▬▬▬▬▬▬▬▬▬▬▬▬▬▬▬▬▬▬▬▬▬

Repair and Remodeling Projects

Some of the more difficult estimates to prepare are for the renovation of existing structures. These projects require the demolition of existing facilities and equipment and then the installation of new materials and equipment. All of this work may have to be done while the existing facility is fully operational. A good example of this might be an inner-city highway project, hospital, or airport. Such structures cannot be totally shut down without significant impact on their surrounding communities, so they must remain in operation while they are repaired or remodeled.

Adding to the complexity of the estimate is the fact that good **"as-built"** documents may not exist for these facilities, requiring that the estimator conduct a thorough inspection of the project and verify actual in-place conditions for the project.

Because many of these projects must be conducted while the facilities remain operational, arrangements must be made to protect the occupants and the surrounding community from significant hardship. Noise, dust, health and safety, access and egress, and debris removal must all be accounted for in the estimate. Much of the work may have to be done at off-peak hours, requiring that labor be paid at a higher hourly rate while working at a lower productivity due to possible extended overtime. Lighting for night operations and police details may have to be budgeted for the work.

As existing facilities are renovated they may need to be brought up to new code requirements, which may require handicap accessibility, asbestos or lead paint removal or encapsulation, or additional fire protection work. These costs can be high, and the construction process can severely hamper existing operations.

A major difficulty associated with renovation projects is accessibility to the project and the fact that work must be done in a sequence that would not occur in new construction. As an example, new foundations may have to be installed under existing structures, requiring special pile driving equipment, shoring, bracing, or underpinning. Equipment installations are also difficult since door openings may not be large enough, requiring special openings to be cut.

As existing materials are demolished the materials must be removed from the site and disposed of. Disposal costs continue to increase—particularly for hazardous materials such as PCBs and asbestos.

As renovation work is conducted it is often important to match existing conditions, particularly in the renovation of historical buildings or other fine architectural work. This can be difficult if matching materials are no longer manufactured or are manufactured at a different size. Paint, tile, and carpet can all be difficult to match and may require work beyond the scope of the initial project.

As the number of good, available building sites continues to decline and our nation's infrastructure erodes, more and more construction projects will be renovations. Estimating renovation is difficult, but if proper attention is paid to the process that will be followed and to the visualization of the environment in which the project will be conducted, a good reliable estimate can be made for the project.

Labor, Material, and Equipment

Once the estimating tasks are identified, categorized, and organized, the team begins the quantity takeoff, which is the foundation of the estimate. The purpose of a quantity takeoff is to accurately determine the quantity of work that needs to be performed on the project. Every work item needs to be measured and quantified using the same units as the pricing guides. Most prices are separated into units of labor, material, and equipment. The quantities of work should be broken out in these same categories. Figure 6–3 highlights a standard entry from the R.S. Means Building Cost Data book, which illustrates the typical organization of cost data.

This book is organized in accordance with the Construction Specifications Institute's 16 Divisions, which is typical for building construction projects. The CSI format uses three sets of numbers to classify items (see Fig. 6–4). Taking the example from Figure 6–3,

092 = lath, plaster and gypsum board

608 = drywall

0150 = ⅜″ thick on walls, standard, no finish included

0200 = the same, except on ceilings

This system of numbering is typical of how most companies involved in the building industry keep their estimating data bases. Companies involved in other industries might set up a different numbering system, but they would still create a hierarchical breakdown. The important factor is that the cost data be readily accessible.

The number in the highlighted line in Figure 6–3 is followed by a written description of the work item. The indentation of the second line indicates that ⅜″ drywall is again used, but now on a ceiling, as the description says. The crew column indicates the configuration of people and equipment that is typically assigned to accomplish this task. In this case two carpenters would be the normal crew used to accomplish the drywall work. Crew sizes are important when using the next column,

092 | Lath, Plaster and Gypsum Board

092 600	Gypsum Board Systems	CREW	DAILY OUTPUT	MAN-HOURS	UNIT	MAT.	LABOR	EQUIP.	TOTAL	TOTAL INCL O&P	
							1994 BARE COSTS				
604 0100	Screwed to grid, channel or joists, 1/2" thick	2 Carp	765	.021	S.F.	.25	.50		.75	1.07	**604**
0200	5/8" thick		765	.021		.27	.50		.77	1.09	
0300	Over 8' high, 1/2" thick		615	.026		.33	.62		.95	1.34	
0400	5/8" thick	↓	615	.026	↓	.35	.62		.97	1.37	
0600	Grid suspension system, direct hung										
0700	1-1/2" C.R.C., with 7/8" hi hat furring channel, 16" O.C.	2 Carp	600	.027	S.F.	.69	.63		1.32	1.77	
0800	24" O.C.		900	.018		.62	.42		1.04	1.35	
0900	3-5/8" C.R.C., with 7/8" hi hat furring channel, 16" O.C.		600	.027		.71	.63		1.34	1.79	
1000	24" O.C.	↓	900	.018	↓	.63	.42		1.05	1.36	
608 0010	DRYWALL Gypsum plasterboard, nailed or screwed to studs,										**608**
0100	unless otherwise noted										
0150	3/8" thick, on walls, standard, no finish included	2 Carp	2,000	.008	S.F.	.15	.19		.34	.47	
0200	On ceilings, standard, no finish included		1,800	.009		.15	.21		.36	.51	
0250	On beams, columns, or soffits, no finish included	↓	675	.024	↓	.24	.56		.80	1.15	
0270											
0300	1/2" thick, on walls, standard, no finish included	2 Carp	2,000	.008	S.F.	.15	.19		.34	.47	
0350	Taped and finished		965	.017		.20	.39		.59	.85	
0400	Fire resistant, no finish included		2,000	.008		.19	.19		.38	.51	
0450	Taped and finished		965	.017		.23	.39		.62	.88	
0500	Water resistant, no finish included		2,000	.008		.23	.19		.42	.55	
0550	Taped and finished		965	.017		.28	.39		.67	.94	
0600	Prefinished, vinyl, clipped to studs	↓	900	.018	↓	.50	.42		.92	1.22	
0650											
1000	On ceilings, standard, no finish included	2 Carp	1,800	.009	S.F.	.16	.21		.37	.52	
1050	Taped and finished		765	.021		.22	.50		.72	1.03	
1100	Fire resistant, no finish included		1,800	.009		.19	.21		.40	.55	
1150	Taped and finished		765	.021		.24	.50		.74	1.05	
1200	Water resistant, no finish included		1,800	.009		.23	.21		.44	.59	
1250	Taped and finished		765	.021		.27	.50		.77	1.09	
1500	On beams, columns, or soffits, standard, no finish included		675	.024		.32	.56		.88	1.24	
1550	Taped and finished		475	.034		.37	.80		1.17	1.68	
1600	Fire resistant, no finish included		675	.024		.29	.56		.85	1.21	
1650	Taped and finished		475	.034		.33	.80		1.13	1.63	
1700	Water resistant, no finish included		675	.024		.34	.56		.90	1.26	
1750	Taped and finished		475	.034		.38	.80		1.18	1.69	
2000	5/8" thick, on walls, standard, no finish included		2,000	.008		.19	.19		.38	.51	
2050	Taped and finished		965	.017		.24	.39		.63	.89	
2100	Fire resistant, no finish included		2,000	.008		.21	.19		.40	.53	
2150	Taped and finished		965	.017		.26	.39		.65	.92	
2200	Water resistant, no finish included		2,000	.008		.28	.19		.47	.61	
2250	Taped and finished		965	.017		.33	.39		.72	.99	
2300	Prefinished, vinyl, clipped to studs		900	.018		.58	.42		1	1.31	
3000	On ceilings, standard, no finish included		1,800	.009		.19	.21		.40	.55	
3050	Taped and finished		765	.021		.24	.50		.74	1.05	
3100	Fire resistant, no finish included		1,800	.009		.21	.21		.42	.57	
3150	Taped and finished		765	.021		.26	.50		.76	1.08	
3200	Water resistant, no finish included		1,800	.009		.28	.21		.49	.65	
3250	Taped and finished		765	.021		.33	.50		.83	1.15	
3500	On beams, columns, or soffits, standard, no finish included		675	.024		.28	.56		.84	1.20	
3550	Taped and finished		475	.034		.34	.80		1.14	1.64	
3600	Fire resistant, no finish included		675	.024		.30	.56		.86	1.22	
3650	Taped and finished		475	.034		.36	.80		1.16	1.67	
3700	Water resistant, no finish included		675	.024		.37	.56		.93	1.30	
3750	Taped and finished		475	.034		.42	.80		1.22	1.73	
4000	Fireproofing, beams or columns, 2 layers, 1/2" thick, incl finish	↓	330	.048	↓	.48	1.15		1.63	2.36	

9 FINISHES

Figure 6–3
Typical Building Construction Cost Data (BCCD) line items, Line Nos. 092-608-0150 and 0200

From *Means Building Construction Cost Data 1995*. Copyright R. S. Means Co., Inc., Kingston, MA, 617-585-7880, all rights reserved.

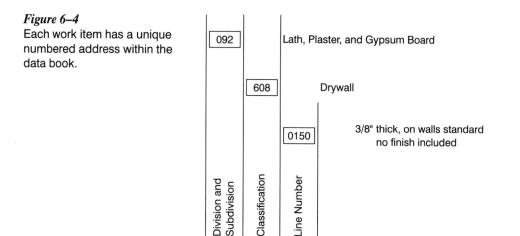

Figure 6–4
Each work item has a unique numbered address within the data book.

092		Lath, Plaster, and Gypsum Board
	608	Drywall
		0150 — 3/8" thick, on walls standard no finish included

Division and Subdivision · Classification · Line Number

daily output. Daily output indicates how many units (sq.ft. of drywall) can be accomplished by the crew (two carpenters) in a day. The manhour column indicates how long, in hours, it will take one carpenter to install 1 sq.ft. of drywall. Daily output and manhours will be covered in further detail in the duration calculation section of Chapter 9.

The key to accomplishing a quantity takeoff correctly is recognizing what units are used to record the cost data for a particular work item. In the drywall example above it is square footage. You can tell by looking at line no. 092-608-0150:

The material cost for the drywall is $0.15/S.F.
The labor cost for the installation is $0.19/S.F.
There is no equipment cost.

In this case the quantity that needs to be calculated for the correct pricing of the drywall is square foot of drywall. If drywall is used on both sides of an 8′ high 100′ long partition, no openings, the quantity of drywall would be 1600 sq.ft. (see Fig. 6–5).

Unlike an assemblies estimate, the drywall is figured separately from the studs and finishes. In the assemblies estimate the unit used is square foot of partition, including all the above components.

The next four columns indicate the price in dollars per unit for material, labor, equipment, and total. The final column is the total column that includes worker fringe benefits, workers' compensation insurance, and installing contractor's overhead and profit. The specifics of these costs will be covered in the next section of this chapter. What is important to remember is that the estimate is priced using the categories of material, labor, and equipment, and to be accomplished correctly the quantities must be calculated using the same units as the unit prices.

Figure 6–6 provides a simple example to illustrate how pricing units are related to quantity units; it takes calculations one step further, determining the actual pricing units and price for a roof covering.

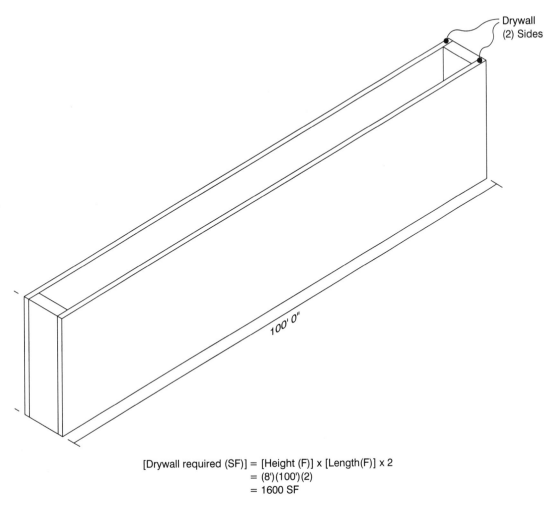

[Drywall required (SF)] = [Height (F)] x [Length(F)] x 2
= (8')(100')(2)
= 1600 SF

Figure 6–5
Square foot of drywall calculation.

Measuring Quantities

The goal of the quantity takeoff process is to calculate every item of the project—no more and no less. To effectively accomplish this, the takeoff must utilize the correct units. When quantifying the project it is also important to think ahead to the scheduling process, as the two are interrelated. The quantities used to price the project also dictate the amount of work required, which will affect duration. Scheduling activities are generally created using visually measurable actions that need to be definable by location and trade. It is therefore important that the quantities be

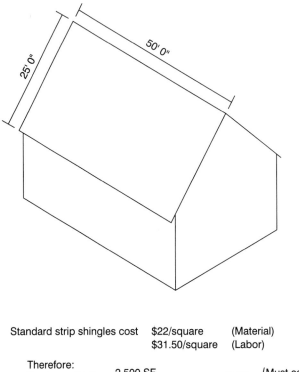

Figure 6–6
Measured units must be the
same as pricing units.

[Roof Area (SF)] = [Length (F)] x [Width(F)] x 2
           = (50')(25')(2)
           = 2,500 SF

Standard strip shingles cost $22/square   (Material)
                                $31.50/square   (Labor)

Therefore:

$$\text{Roof area (SQ)} = \frac{2{,}500 \text{ SF}}{100 \text{ SF/SQ}} = 25 \text{ SQ} \quad \left(\begin{array}{l}\text{Must convert} \\ \text{from SF to SQ}\end{array}\right)$$

Material cost = (25 SQ)($22/SQ)   = $550
Labor cost   = (25 SQ)($31.50/SQ) = $787.50

dividable into these elements (see Fig. 6–7). This section will describe some of the techniques that are utilized when accomplishing a quantity takeoff.

The use of preprinted forms is a common tool used to help coordinate the take-off process, especially if more than one person is involved. Also if the project being estimated is similar to a past project, these forms can be set up with all the estimate items identified, thereby serving as a checklist for the estimator (see Fig. 6–8).

Approach each section of the project in as orderly a fashion as possible. Use printed dimensions whenever possible and add up dimensions when possible to utilize a single entry. The goal is to minimize the amount of measurements and calculations that need to be made. When measuring dimensions, ensure that the correct scale is being used. Be careful that the drawings have not been reduced, or that the

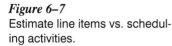

Figure 6–7
Estimate line items vs. scheduling activities.

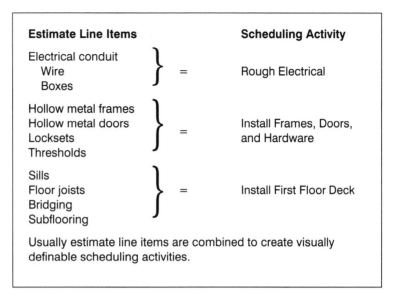

Estimate Line Items		Scheduling Activity

Electrical conduit
    Wire
    Boxes
	} =	Rough Electrical

Hollow metal frames Hollow metal doors Locksets Thresholds	} =	Install Frames, Doors, and Hardware
Sills Floor joists Bridging Subflooring	} =	Install First Floor Deck

Usually estimate line items are combined to create visually definable scheduling activities.

area being worked in is not of a different scale or NTS (not to scale). Whenever possible take advantage of repeated project elements such as multiple floors, elevations, or interior partitions (see Fig. 6–9).

It is a good idea when working through a drawing to mark the drawing using pencil to note what has and has not been included. On the preprinted forms make notes, sketches, or whatever is necessary to make the checking process easier.

When performing calculations, avoid rounding off until the final quantity summary. Work on the front page only, as work on the back of pages can easily be lost or forgotten. Be consistent when listing dimensions; for example, specify length × width × height in that order, and maintain that order. Work in pencil, and put all the figures in the correct columns on the preprinted forms.

The quantities calculated must be adjusted for waste, as well as soil shrinkage and swell. In every construction project more material must be purchased than is actually needed per the drawings. As an example, when ordering concrete, some of the concrete may be spilled when transported from the truck by bucket or wheelbarrow. In wood framing the joists, studs, and plywood are all bought in designated lengths, but cut down on site to fit a specific dimension. Waste occurs in tilework, wallpaper, and paint. The only material exceptions are custom building elements such as windows, precast panels, structural steel, or compressors. The uniqueness of the project, transportation, or sometimes misplacement and vandalism causes waste (see Fig. 6–10). In some instances, soil being a good example, the volume of the material that must be removed and transported is greater than the quantity measured in the ground; this is called **swell.** Conversely, when backfilling, the soil volume when delivered and compacted sometimes is less than what was brought to the

Figure 6–8
Preprinted estimate form.
From *Means Form Book*. Copyright R. S. Means Co., Inc., Kingston, MA, 617-585-7880, all rights reserved.

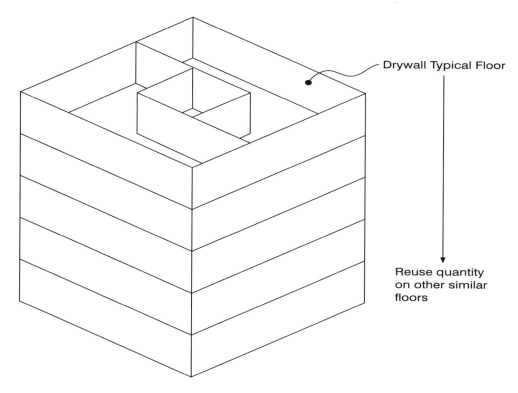

Drywall Typical Floor

Reuse quantity on other similar floors

As illustrated, repeated elements, such as drywall on one floor of a commercial building, can be taken off once and reused on other typical floors.

Figure 6–9
Take advantage of building symmetry.

Material	Typical Adjustment	Reasons for Waste
Steel reinforcement	10%	For splices and corners
Welded wire fabric	5%	For overlapping
Concrete	7%	Waste/spillage and shrinkage
Concrete block	2%	Waste/breakage and trim
Wall sheathing	5%	Cut to fit
Drywall	5%	Cut to fit
Resilient flooring	5%	Cut to fit
Tile	5%	Waste/breakage and trim

Figure 6–10
Typical quantity adjustments for residential construction.

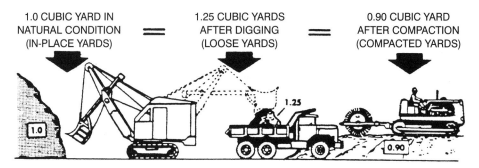

COMMON EARTH

Bank Measure:	Volume of earth in natural state before loosening
Loose Measure:	Volume of earth as transported in trucks
Compacted Measure:	Volume of earth after placed and compacted

Soil Type	Bank	Loose	Compacted
Clay	1.00	1.27	0.90
Earth	1.00	1.25	0.90
Sand	1.00	1.12	0.95

Figure 6–11
Shrink and swell illustration.
From *Means Estimating Handbook 1990*. Copyright R. S. Means Co., Inc., Kingston, MA, 617-585-7880, all rights reserved.

site; this is called **shrinkage.** The amount of swell and shrinkage that occurs is a factor of the type of material being handled and must be adjusted for in the quantity takeoff (see Fig. 6–11).

The quantity takeoff process requires a strong understanding of the work involved in each of the different disciplines of a project. For example, when taking off structural steel work, the estimator visualizes the work and quantifies all the elements of the activity. The number of structural steel pieces are counted by type of steel, with paint requirements, fabrication requirements, and testing and installation requirements figured into each type. The same level of detail needs to be considered in each of the work areas of the project. Often the best approach is to build the project item by item on the takeoff form, then quantify each item. The use of preprinted forms can be a real help at this point. Items that require a price from a vendor or subcontractor should be noted on the estimating form with an asterisk to ensure they are not forgotten.

In summary, the quantity takeoff process requires an organized approach to every drawing and building element to ensure that not a single work item is missed. The estimator works drawing by drawing, with corresponding specification pages, floor by floor, from plans to elevations to sections to details, marking the drawings while proceeding.

Use of abbreviations, conversion of feet and inches, and shortcuts such as using design symmetry (repeating common project elements) all save time and improve accuracy, but care should be taken to cover each item, or a costly mistake can occur.

Unit Pricing

With the quantity takeoff complete, the estimator has determined all the typical components that go into the project and how many of each type are necessary. The next task is for the estimating team to determine how much each unit will cost to produce, deliver to the site, accept and store at the site, install in the correct position, and maintain until the project is accepted. The production of the product and the delivery to the site is included in the material unit price. The labor unit price includes the crew cost involved in the installation of the material at the job site. The equipment unit price covers the cost of the equipment necessary to install the material. Project overhead will cover the costs of accepting the material, storing it at the job site, and protecting the work until the project is accepted. Overall corporate overhead will include the costs of preparing the estimate, marketing the company, and providing broadbased technical and administrative support to the project. This section of the chapter will cover the derivation and use of the material, labor, and equipment unit prices. Project overhead, home office overhead, taxes, bonding, and profit will be covered in the last section of the chapter.

Pricing Sources

The cost data used in this section are taken from the R.S. Means Building Cost Data book, which is publicly available to cost estimators. In-house unit price data—just like conceptual, square foot, and assemblies cost data—is often developed and maintained by the larger design and construction companies. However, for those companies who do not have the resources to develop and maintain their own data base, or if a price is necessary which is not available in their own data base, the Means data provide a reliable alternative. Means data, government reports, cost reporting services, and technical press reports all compile information on past costs and project future costs and trends. This information can be good, but should be checked for accuracy as well as for the region and industry being considered.

The figures provided by cost reports and books like Means are projections of costs and are not meant to be the basis of a contract which guarantees the cost of the work. The best and most reliable costs that an estimator can get are those provided by a supplier or a subcontractor. The costs quoted are usually good for a specific period of time and for a certain amount of material or work. Subcontractor quotes or bids may be required to be accompanied by a bid bond which guarantees that if accepted a contract will be signed for the amount quoted. When using quotes an estimator needs to verify that the price quoted covers the entire scope of the work. Any miscommunications could lead to serious financial or contractual problems.

There are many other sources to investigate for pricing. If the work is being done using a union work force, then wage rates, fringe benefits, and productivity rates are published by the local union. Local and state government offices also publish tax rates, fees, permits required, and social security and unemployment insurance rates. Equipment rental rates are available from local rental agencies. Quotes can also be received from insurance and bonding providers. For all these costs, it is important to get a guarantee as to the rate that will be charged at the time of construction. If this is not possible, the costs will have to be projected and escalated to the future.

Material Costs

Of all the prices that need to be identified, the project materials prices are generally the easiest to determine. The most reliable source is the supplier. Published prices in catalogs are also a good source, or prices can be obtained from previous projects or published unit price books, such as Means.

When receiving a price quote or when looking up a material unit price, it is important to consider the following questions:

1. Is the material quoted the actual item specified? For example, is it the correct model number, color, and finish?
2. Is the quoted price valid until the scheduled time of delivery?
3. Does the price include delivery to the job site?
4. Are adequate warranties and guarantees being provided?
5. What is the lead time to delivery?
6. Does the supplier maintain adequate stock?
7. What are the payment terms? Are there discount or credit options, and so forth?
8. How reputable is the supplier?

Note in the Means drywall example above the $0.15/sq.ft. material price includes the price of manufacturing and delivering the drywall to the job site. The price does not include any overhead, profit, or sales tax.

Labor Costs

The pricing of labor is the most difficult factor to determine. In pricing labor, the estimator figures two different components: the hourly wage rate, and the crew productivity.

Wage Rates

The wage rate is figured by the trade(s) involved and the rates paid. Union rates are available from the union locals, employer bargaining groups, or from publications such as Means. Nonunion labor rates are determined by each company and are dependent on the geographic area. In either case, wage escalation needs to be factored in, particularly on long duration projects. Union rates are generally negotiated for 1–3 year periods by trade, so these agreements will have to be researched. Some long duration projects have escalation clauses that protect the bidders, but on projects where this is not the case the estimator will have to estimate wage escalation into the future.

Each year Means publishes union labor rates by trade on the inside back cover of the Building Construction Cost Data book. The wages published are the average wage rates of the 30 major United States cities (see Fig. 6–12). These are the wage rates that are used throughout the data book.

There are, however, conditions on the job that may increase these rates. For projects with an aggressive schedule, overtime should be factored in, and if the project is located in a remote area or involves hazardous conditions, the average labor rates will be higher.

Productivity

Once the quantity of work is known and the hourly wage determined, the last step in determining the labor cost of the project is to estimate how long the activity will take. The formula for labor cost is as follows:

Labor cost of activity = Labor rate × Activity duration

Determining crew durations or productivity requires experience and the ability to visualize how the work will be done in the field. The information an estimator needs to know includes the answers to questions such as the following:

1. Will the crew be operating at full efficiency?
2. Is other work occurring concurrently that will interfere?

Installing Contractor's Overhead & Profit

Below are the **average** installing contractor's percentage mark-ups applied to base labor rates to arrive at typical billing rates.

Column A: Labor rates are based on union wages averaged for 30 major U.S. cities. Base rates including fringe benefits are listed hourly and daily. These figures are the sum of the wage rate and employer-paid fringe benefits such as vacation pay, employer-paid health and welfare costs, pension costs, plus appropriate training and industry advancement funds costs.

Column B: Workers' Compensation rates are the national average of state rates established for each trade.

Column C: Column C lists average fixed overhead figures for all trades. Included are Federal and State Unemployment costs set at 7.3%; Social Security Taxes (FICA) set at 7.65%; Builder's Risk Insurance costs set at 0.34%; and Public Liability costs set at 1.55%. All the percentages except those for Social Security Taxes vary from state to state as well as from company to company.

Columns D and E: Percentages in Columns D and E are based on the presumption that the installing contractor has annual billing of $500,000 and up. Overhead percentages may increase with smaller annual billing. The overhead percentages for any given contractor may vary greatly and depend on a number of factors, such as the contractor's annual volume, engineering and logistical support costs, and staff requirements. The figures for overhead and profit will also vary depending on the type of job, the job location, and the prevailing economic conditions. All factors should be examined very carefully for each job.

Column F: Column F lists the total of Columns B, C, D, and E.

Column G: Column G is Column A (hourly base labor rate) multiplied by the percentage in Column F (O&P percentage).

Column H: Column H is the total of Column A (hourly base labor rate) plus Column G (Total O&P).

Column I: Column I is Column H multiplied by eight hours.

		A		B	C	D	E	F		G	H		I
		Base Rate Incl. Fringes		Workers' Comp. Ins.	Average Fixed Overhead	Overhead	Profit	Total Overhead & Profit			Rate with O & P		
Abbr.	Trade	Hourly	Daily					%	Amount		Hourly		Daily
Skwk	Skilled Workers Average (35 trades)	$24.65	$197.20	19.0%	16.8%	13.0%	10.0%	58.8%	$14.50		$39.15		$313.20
	Helpers Average (5 trades)	18.60	148.80	20.0		11.0		57.8	10.75		29.35		234.80
	Foreman Average, Inside ($.50 over trade)	25.15	201.20	19.0		13.0		58.8	14.75		39.90		319.20
	Foreman Average, Outside ($2.00 over trade)	26.65	213.20	19.0		13.0		58.8	15.65		42.30		338.40
Clab	Common Building Laborers	19.00	152.00	20.6		11.0		58.4	11.10		30.10		240.80
Asbe	Asbestos Workers	26.90	215.20	18.1		16.0		60.9	16.40		43.30		346.40
Boil	Boilermakers	28.05	224.40	11.3		16.0		54.1	15.20		43.25		346.00
Bric	Bricklayers	24.55	196.40	18.2		11.0		56.0	13.75		38.30		306.40
Brhe	Bricklayer Helpers	19.50	156.00	18.2		11.0		56.0	10.90		30.40		243.20
Carp	Carpenters	23.80	190.40	20.6		11.0		58.4	13.90		37.70		301.60
Cefi	Cement Finishers	23.25	186.00	12.0		11.0		49.8	11.60		34.85		278.80
Elec	Electricians	27.50	220.00	7.6		16.0		50.4	13.85		41.35		330.80
Elev	Elevator Constructors	28.15	225.20	9.0		16.0		51.8	14.60		42.75		342.00
Eqhv	Equipment Operators, Crane or Shovel	25.40	203.20	12.6		14.0		53.4	13.55		38.95		311.60
Eqmd	Equipment Operators, Medium Equipment	24.35	194.80	12.6		14.0		53.4	13.00		37.35		298.80
Eqlt	Equipment Operators, Light Equipment	23.40	187.20	12.6		14.0		53.4	12.50		35.90		287.20
Eqol	Equipment Operators, Oilers	20.75	166.00	12.6		14.0		53.4	11.10		31.85		254.80
Eqmm	Equipment Operators, Master Mechanics	25.95	207.60	12.6		14.0		53.4	13.85		39.80		318.40
Glaz	Glaziers	23.80	190.40	15.4		11.0		53.2	12.65		36.45		291.60
Lath	Lathers	23.70	189.60	13.3		11.0		51.1	12.10		35.80		286.40
Marb	Marble Setters	24.65	197.20	18.2		11.0		56.0	13.80		38.45		307.60
Mill	Millwrights	25.10	200.80	12.6		11.0		50.4	12.65		37.75		302.00
Mstz	Mosaic and Terrazzo Workers	24.20	193.60	10.4		11.0		48.2	11.65		35.85		286.80
Pord	Painters, Ordinary	22.20	177.60	15.8		11.0		53.6	11.90		34.10		272.80
Psst	Painters, Structural Steel	23.10	184.80	60.9		11.0		98.7	22.80		45.90		367.20
Pape	Paper Hangers	22.40	179.20	15.8		11.0		53.6	12.00		34.40		275.20
Pile	Pile Drivers	23.95	191.60	31.5		16.0		74.3	17.80		41.75		334.00
Plas	Plasterers	23.30	186.40	16.6		11.0		54.4	12.70		36.00		288.00
Plah	Plasterer Helpers	19.75	158.00	16.6		11.0		54.4	10.75		30.50		244.00
Plum	Plumbers	28.30	226.40	9.6		16.0		52.4	14.85		43.15		345.20
Rodm	Rodmen (Reinforcing)	26.40	211.20	36.2		14.0		77.0	20.35		46.75		374.00
Rofc	Roofers, Composition	21.55	172.40	34.2		11.0		72.0	15.50		37.05		296.40
Rots	Roofers, Tile and Slate	21.60	172.80	34.2		11.0		72.0	15.55		37.15		297.20
Rohe	Roofer Helpers (Composition)	15.35	122.80	34.2		11.0		72.0	11.05		26.40		211.20
Shee	Sheet Metal Workers	27.35	218.80	13.2		16.0		56.0	15.30		42.65		341.20
Spri	Sprinkler Installers	30.35	242.80	9.9		16.0		52.7	16.00		46.35		370.80
Stpi	Steamfitters or Pipefitters	28.30	226.40	9.6		16.0		52.4	14.85		43.15		345.20
Ston	Stone Masons	24.70	197.60	18.2		11.0		56.0	13.85		38.55		308.40
Sswk	Structural Steel Workers	26.50	212.00	42.7		14.0		83.5	22.15		48.65		389.20
Tilf	Tile Layers (Floor)	24.00	192.00	10.4		11.0		48.2	11.55		35.55		284.40
Tilh	Tile Layer Helpers	19.25	154.00	10.4		11.0		48.2	9.30		28.55		228.40
Trlt	Truck Drivers, Light	19.40	155.20	16.3		11.0		54.1	10.50		29.90		239.20
Trhv	Truck Drivers, Heavy	19.70	157.60	16.3		11.0		54.1	10.65		30.35		242.80
Sswl	Welders, Structural Steel	26.50	212.00	42.7		14.0		83.5	22.15		48.65		389.20
Wrck	*Wrecking	19.00	152.00	42.5	▼	11.0	▼	80.3	15.25		34.25		274.00

*Not included in Averages.

Figure 6–12

Union labor rates by trade, and contractor's overhead and profit.

From *Means Building Construction Cost Data 1995*. Copyright R. S. Means Co., Inc., Kingston, MA, 617-585-7880, all rights reserved.

3. What weather conditions are to be expected?

4. Will the crew be working on ladders or scaffolding or on the ground?

5. Will extended periods of overtime be occurring?

Past project experience is essential in answering these questions. Because of this need, in most project offices the quantity takeoff work is done by the junior estimators and the pricing by the senior estimators.

In the drywall example the labor cost of $0.19/sq.ft. is calculated in the following manner:

Time required for one carpenter to install 1 square foot of drywall = .008 hours

Base rate for one carpenter = $23.80/hour

Therefore:

$$\text{Labor cost per sq.ft.} = (.008 \text{ hr} \times \$23.80) \text{ per sq.ft.}$$
$$= \$0.19 \text{ per sq.ft.}$$

Equipment Costs

Equipment costs are the last element needed before assembling the estimate. These costs are of two general types: the equipment itself and the cost of operating it. Under the equipment would be the cost of ownership, lease, or rental. This cost would cover interest, storage, insurance, taxes, and license. If the equipment used is owned by the company, these costs could be determined by talking to the company's financial people. If the equipment is to be leased or rented, the equipment supplier can provide a written quote.

The second cost is that of operating the equipment for as long as it is needed. This would include the cost of gasoline, oil, periodic maintenance, transportation, and mobilization. The cost of an operator should be covered in the labor line item, but this must be verified. Figure 6–13 provides an illustration. It computes the daily costs for a B76 crew which utilizes a 50-ton crawler crane.

Equipment costs can be figured on an item-by-item basis, or can be covered on a project basis in general conditions or project overhead. As an example, formwork requires the use of small power tools. Looking at the line item in the equipment (Equip) column, it is covered at the unit cost of $0.07/SFCA (see Fig. 6–14).

The cost of a tower crane (3000 lb, 150 ft) which is used by all trades, however, could be covered on a project basis (see Fig. 6–15). The crane would probably be rented on a monthly basis of $6,700 per month and show an operating cost of $10.80 per hour.

Equipment, labor, and materials make up the three major cost categories of a project. When researching prices it is necessary to predict what the cost will be at

Crew B-76	Hr.	Daily	Hr.	Daily	Bare Costs	Incl. O&P
1 Dock Builder Foreman	$25.95	$207.60	$45.25	$362.00	$24.14	$40.42
5 Dock Builders	23.95	958.00	41.75	1670.00		
2 Equip. Oper. (crane)	25.40	406.40	38.95	623.20		
1 Equip. Oper. Oiler	20.75	166.00	31.85	254.80		
1 Crawler Crane, 50 Ton		824.00		906.40		
1 Barge, 400 Ton		416.60		458.25		
1 Hammer, 15K Ft. Lbs.		282.20		310.40		
60 L.F. Leads, 15K Ft. Lbs.		60.00		66.00		
1 Air Compr., 600 C.F.M.		265.00		291.50		
2-50 Ft. Air Hoses, 3″ Dia.		29.60		32.55	26.08	28.68
72 M. H., Daily Totals		$3615.40		$4975.10	$50.22	$69.10

Referencing the dock building crew (B76) above, let's look at the 50-ton crawler crane. The daily cost of $824 is taken from line 016–460–1000 listed below.

		Hr. Oper. Cost	Rent per Day	Rent per Week	Rent per Month	Crew Equip. Cost
0600	Crawler, cable, 1/2 C.Y., 15 tons at 12′ radius	17.30	450	1,355	4,075	409.40
0700	3/4 C.Y., 20 tons at 12′ radius	17.90	490	1,465	4,400	436.20
0800	1 C.Y., 25 tons at 12′ radius	19.25	530	1,590	4,775	472
0900	1 1/2 C.Y., 40 tons at 12′ radius	27.85	815	2,450	7,350	712.80
1000	2 C.Y., 50 tons at 12′ radius	32.50	940	2,820	8,450	824
1100	3 C.Y., 75 tons at 12′ radius	39.80	805	2,410	7,225	800.40

The $824 crew equipment cost used in the B76 crew above is calculated as follows:

Crew equipment cost = (Rent per week/5)+[(8) Hourly operating cost]
  = (2,820/5)+[(8)(32.50)]
  = $564 + $260
  = $824
Equipment cost with O&P = $824 + 10%
  = $906.40

Figure 6–13
Crew equipment analysis.
From *Means Building Construction Cost Data 1995*. Copyright R. S. Means Co., Inc., Kingston, MA, 617-585-7880, all rights reserved.

031 100 | Struct C.I.P. Formwork

		Description	CREW	DAILY OUTPUT	MAN-HOURS	UNIT	MAT.	LABOR	EQUIP.	TOTAL	TOTAL INCL O&P	
							1994 BARE COSTS					
150	6650	4 use	C-1	225	.142	SFCA	.57	3.21	.12	3.90	5.85	150
	7000	Edge forms to 6" high, on elevated slab, 4 use		500	.064	L.F.	.31	1.45	.06	1.82	2.70	
	7100	7" to 12" high, 4 use		350	.091	SFCA	.54	2.07	.08	2.69	3.95	
	7500	Depressed area forms to 12" high, 4 use		300	.107	L.F.	.56	2.41	.09	3.06	4.53	
	7550	12" to 24" high, 4 use		175	.183		.76	4.13	.16	5.05	7.55	
	8000	Perimeter deck and rail for elevated slabs, straight		90	.356		7.35	8.05	.31	15.71	21	
	8050	Curved		65	.492		10.10	11.15	.42	21.67	29	
	8500	Void forms, round fiber, 3" diameter		450	.071		.41	1.61	.06	2.08	3.07	
	8550	4" diameter, void		425	.075		.44	1.70	.06	2.20	3.25	
	8600	6" diameter, void		400	.080		.77	1.81	.07	2.65	3.79	
	8650	8" diameter, void		375	.085		1.26	1.93	.07	3.26	4.52	
	8700	10" diameter, void		350	.091		2.11	2.07	.08	4.26	5.70	
	8750	12" diameter, void		300	.107		2.78	2.41	.09	5.28	7	
	8800	Metal end closures, loose, minimum				C	26			26	28.50	
	8850	Maximum				"	142			142	156	
154	0010	FORMS IN PLACE, EQUIPMENT FOUNDATIONS 1 use	C-2	160	.300	SFCA	1.96	7	.23	9.19	13.50	154
	0050	2 use		190	.253		1.07	5.90	.19	7.16	10.75	
	0100	3 use		200	.240		.86	5.60	.18	6.64	10	
	0150	4 use		205	.234		.69	5.45	.18	6.32	9.60	
158	0010	FORMS IN PLACE, FOOTINGS Continuous wall, 1 use	C-1	375	.085	SFCA	1.19	1.93	.07	3.19	4.44	158
	0050	2 use		440	.073		.65	1.64	.06	2.35	3.38	
	0100	3 use		470	.068		.48	1.54	.06	2.08	3.02	
	0150	4 use		485	.066		.38	1.49	.06	1.93	2.84	
	0500	Dowel supports for footings or beams, 1 use		500	.064	L.F.	.53	1.45	.06	2.04	2.93	
	1000	Integral starter wall, to 4" high, 1 use		400	.080		1.03	1.81	.07	2.91	4.07	
	1500	Keyway, 4 use, tapered wood, 2" x 4"	1 Carp	530	.015		.16	.36		.52	.75	
	1550	2" x 6"		500	.016		.22	.38		.60	.85	
	2000	Tapered plastic, 2" x 3"		530	.015		.43	.36		.79	1.05	
	2050	2" x 4"		500	.016		.52	.38		.90	1.17	
	2250	For keyway hung from supports, add		150	.053		.66	1.27		1.93	2.73	
	2260											
	3000	Pile cap, square or rectangular, 1 use	C-1	290	.110	SFCA	1.84	2.49	.09	4.42	6.10	
	3050	2 use		346	.092		1.04	2.09	.08	3.21	4.54	
	3100	3 use		371	.086		.77	1.95	.07	2.79	4.02	
	3150	4 use		383	.084		.68	1.89	.07	2.64	3.82	
	4000	Triangular or hexagonal caps, 1 use		225	.142		1.82	3.21	.12	5.15	7.25	
	4050	2 use		280	.114		1.03	2.58	.10	3.71	5.35	
	4100	3 use		305	.105		.77	2.37	.09	3.23	4.71	
	4150	4 use		315	.102		.64	2.30	.09	3.03	4.44	
	5000	Spread footings, 1 use		305	.105		1.42	2.38	.09	3.89	5.45	
	5050	2 use		371	.086		.91	1.95	.07	2.93	4.17	
	5100	3 use		401	.080		.68	1.80	.07	2.55	3.69	
	5150	4 use		414	.077		.51	1.75	.07	2.33	3.40	
	6000	Supports for dowels, plinths or templates, 2' x 2'		25	1.280	Ea.	2.60	29	1.10	32.70	50	
	6050	4' x 4' footing		22	1.455		5.60	33	1.25	39.85	59.50	
	6100	8' x 8' footing		20	1.600		11.35	36	1.38	48.73	71.50	
	6150	12' x 12' footing		17	1.882		18.50	42.50	1.62	62.62	90	
	7000	Plinths, 1 use		250	.128	SFCA	2.02	2.89	.11	5.02	6.95	
	7100	4 use		270	.119		.56	2.68	.10	3.34	4.97	
162	0010	FORMS IN PLACE, GRADE BEAM 1 use	C-2	530	.091		1.36	2.11	.07	3.54	4.92	162
	0050	2 use		580	.083		.82	1.93	.06	2.81	4.03	
	0100	3 use		600	.080		.62	1.87	.06	2.55	3.71	
	0150	4 use		605	.079		.52	1.85	.06	2.43	3.57	
166	0010	FORMS IN PLACE, MAT FOUNDATION 1 use		290	.166	SFCA	1.49	3.86	.13	5.48	7.90	166
	0050	2 use		310	.155		.85	3.61	.12	4.58	6.75	

Figure 6–14

Costs of formwork.

From *Means Building Construction Cost Data 1995*. Copyright R. S. Means Co., Inc., Kingston, MA, 617-585-7880, all rights reserved.

016 400 | Equipment Rental

		UNIT	HOURLY OPER. COST	RENT PER DAY	RENT PER WEEK	RENT PER MONTH	CREW EQUIPMENT COST	
460	1200	100 ton capacity, standard boom	Ea.	37.90	1,400	4,200	12,600	1,143	**460**
	1300	165 ton capacity, standard boom		59.40	2,225	6,700	20,100	1,815	
	1400	200 ton capacity, 150' boom		112.65	2,450	7,350	22,100	2,371	
	1500	450' boom		127	3,075	9,250	27,800	2,866	
	1600	Truck mounted, cable operated, 6 x 4, 20 tons at 10' radius		12.35	655	1,960	5,875	490.80	
	1700	25 tons at 10' radius		18.95	1,050	3,175	9,525	786.60	
	1800	8 x 4, 30 tons at 10' radius		25.90	605	1,820	5,450	571.20	
	1900	40 tons at 12' radius		27.15	740	2,215	6,650	660.20	
	2000	8 x 4, 60 tons at 15' radius		41.10	885	2,660	7,975	860.80	
	2050	82 tons at 15' radius		32.15	1,625	4,850	14,600	1,227	
	2100	90 tons at 15' radius		44.55	1,025	3,065	9,200	969.40	
	2200	115 tons at 15' radius		46.40	1,850	5,550	16,700	1,481	
	2300	150 tons at 18' radius		69.35	1,550	4,635	13,900	1,482	
	2350	165 tons at 18' radius		70.65	2,150	6,450	19,400	1,855	
	2400	Truck mounted, hydraulic, 12 ton capacity		20.75	400	1,200	3,600	406	
	2500	25 ton capacity		21.35	535	1,605	4,825	491.80	
	2550	33 ton capacity		22.10	805	2,410	7,225	658.80	
	2600	55 ton capacity		31.05	800	2,405	7,225	729.40	
	2700	80 ton capacity		33.95	1,200	3,585	10,800	988.60	
	2800	Self-propelled, 4 x 4, with telescoping boom, 5 ton		7.90	240	720	2,150	207.20	
	2900	12-1/2 ton capacity		14.90	555	1,670	5,000	453.20	
	3000	15 ton capacity		16.75	450	1,350	4,050	404	
	3100	25 ton capacity		19.10	665	2,000	6,000	552.80	
	3200	Derricks, guy, 20 ton capacity, 60' boom, 75' mast		8.40	270	810	2,425	229.20	
	3300	100' boom, 115' mast		15.40	485	1,450	4,350	413.20	
	3400	Stiffleg, 20 ton capacity, 70' boom, 37' mast		10.50	350	1,050	3,150	294	
	3500	100' boom, 47' mast		16.65	595	1,785	5,350	490.20	
	3550	Helicopter, small, lift to 1250 lbs. maximum		233.80	2,350	7,030	21,100	3,276	
	3600	Hoists, chain type, overhead, manual, 3/4 ton		.06	5	15	45	3.50	
	3900	10 ton		.25	23.50	70	210	16	
	4000	Hoist and tower, 5000 lb. cap., portable electric, 40' high		3.79	157	470	1,400	124.30	
	4100	For each added 10' section, add			7.35	22	66	4.40	
	4200	Hoist and single tubular tower, 5000 lb. electric, 100' high		5.12	215	645	1,925	169.95	
	4300	For each added 6'-6" section, add		.61	18	54	162	15.70	
	4400	Hoist and double tubular tower, 5000 lb., 100' high		5.40	233	700	2,100	183.20	
	4500	For each added 6'-6" section, add		.06	11.35	34	102	7.30	
	4550	Hoist and tower, mast type, 6000 lb., 100' high		5	253	760	2,275	192	
	4570	For each added 10' section, add		.12	8.55	25.70	77	6.10	
	4600	Hoist and tower, personnel, electric, 2000 lb., 100' @ 125 FPM		9.42	635	1,900	5,700	455.35	
	4700	3000 lb., 100' @ 200 FPM		10.10	690	2,070	6,200	494.80	
	4800	3000 lb., 150' @ 300 FPM		10.80	745	2,230	6,700	532.40	
	4900	4000 lb., 100' @ 300 FPM		11.50	795	2,390	7,175	570	
	5000	6000 lb., 100' @ 275 FPM		12.20	850	2,550	7,650	607.60	
	5100	For added heights up to 500', add	L.F.		1.10	3.30	9.90	.65	
	5200	Jacks, hydraulic, 20 ton	Ea.	.13	2.10	6.30	18.90	2.30	
	5500	100 ton	"	.15	18.35	55	165	12.20	
	6000	Jacks, hydraulic, climbing with 50' jackrods							
	6010	and control consoles, minimum 3 mo. rental							
	6100	30 ton capacity	Ea.	.05	96.50	290	870	58.40	
	6150	For each added 10' jackrod section, add			2.03	6.10	18.30	1.20	
	6300	50 ton capacity			157	470	1,400	94	
	6350	For each added 10' jackrod section, add			3.05	9.15	27.50	1.85	
	6500	125 ton capacity			450	1,350	4,050	270	
	6550	For each added 10' jackrod section, add			22	66	198	13.20	
	6600	Cable jack, 10 ton capacity with 200' cable			76.50	230	690	46	
	6650	For each added 50' of cable, add			4.58	13.75	41.50	2.75	
490	0010	WELLPOINT EQUIPMENT RENTAL See also division 021-444 R021 -440							**490**
	0020	Based on 2 months rental							

Figure 6–15

Equipment rental costs.

From *Means Building Construction Cost Data 1995*. Copyright R. S. Means Co., Inc., Kingston, MA, 617-585-7880, all rights reserved.

150

the time that the work will be done. When possible, the estimator should try to "lock in," or guarantee, the price if possible. Particularly in the case of labor and equipment, productivity is a factor since the length of time the worker or equipment is on the job site affects the final cost of the activity.

Estimate Setup

Most estimates are compiled by a number of people and organizations. Putting the prices together into a coherent final report requires consideration of many numbers from many different sources. To this point, the estimate has been prepared working within specific divisions or trades. The task now is to add overhead, profit, taxes, and escalation if applicable.

Format

To accomplish this, the first step is to bring the costs of each specialty area "forward" to an estimate summary page. As can be seen in the summary page in Figure 6–16, the estimate is broken down by the 16 CSI divisions (item 1 in that figure, which shows sample calculations for a rather large college student union project). This is common practice for building construction projects. The estimate can be summarized in whatever breakdown form is used. The important factor is to be able to separate subcontractor from in-house work. In-house work needs to be separated further into material, labor, and equipment.

The subcontractor price will include appropriate tax, insurance, and overhead and profit for the subcontractor. Therefore, the subcontractor total will have to be adjusted only for applicable general contractor overhead and profit. In-house labor, equipment, and material costs, however, will need to have overhead and profit added, too. Also, material and equipment prices are taxed in most states (item 2 in Fig. 6–16, at 5%), whereas labor is not. The overhead markup on labor is much higher than it is on material and equipment. Separating these categories allows the proper adjustments to these costs.

In the Means Condensed Estimate Summary page, Division 1 is General Requirements. This line item is also called project overhead. It picks up the costs associated with operating the job site in the field. These costs would include the cost of the field office people, safety, security, photography, and cleanup. These costs are typically itemized with quantities and unit prices figured exactly. Understanding project duration is important to accurately figure the project's general requirements cost. See Figure 6–17 for a list of typical general requirement/project overhead items.

CONDENSED ESTIMATE SUMMARY

SHEET NO.

PROJECT: COLLEGE STUDENT UNION ESTIMATE NO.

LOCATION TOTAL AREA/VOLUME DATE

ARCHITECT COST PER S.F./C.F. NO. OF STORIES: 2

PRICES BY: EXTENSIONS BY: CHECKED BY:

DIV.	DESCRIPTION	MATERIAL	LABOR	EQUIPMENT	SUBCONTRACT	TOTAL
1.0	General Requirements	26000	45000	22600		93500
2.0	Site Work				80500	80500
3.0	Concrete				373400	373400
4.0	Masonry	63523	44002	14500		122025
5.0	Metals	29500	8500	2500		35500
6.0	Carpentry	28750	19915			48665
7.0	Moisture & Thermal Protection				42750	42750
8.0	Doors, Windows, Glass	66023	45734	4500		116257
9.0	Finishes	65545	68104			133649
10.0	Specialties	20400	13600			34000
11.0	Equipment	28500	9500			38000
12.0	Furnishings					
13.0	Special Construction					
14.0	Conveying Systems				143500	143500
15.0	Mechanical				312000	312000
16.0	Electrical				217500	217500
	Subtotals	323241	254355	44000	1167650	1791246
(2)	Sales Tax 5%	16162		2200		18362
(3)	Overhead & Profit % MAT 10% LABOR 58% EQUIP 10% SUBCONTRACT 10%	32324	149961	4400	116765	303250
	Subtotal	371727	403916	50600	1286615	2112858
	Profit %					
(4)	Contingency 3%					63386
(5)	Adjustments					
	TOTAL BID					2176244

Figure 6–16
Estimate summary page.

PROJECT OVERHEAD SUMMARY

	SHEET NO.
PROJECT	ESTIMATE NO.

LOCATION	ARCHITECT	DATE

QUANTITIES BY:	PRICES BY:	EXTENSIONS BY:	CHECKED BY:

DESCRIPTION							
Job Organization: Superintendent							
Project Manager							
Timekeeper & Material Clerk							
Clerical							
Safety, Watchman & First Aid							
Travel Expense: Superintendent							
Project Manager							
Engineering: Layout							
Inspection/Quantities							
Drawings							
CPM Schedule							
Testing: Soil							
Materials							
Structural							
Equipment: Cranes							
Concrete Pump, Conveyor, Etc.							
Elevators, Hoists							
Freight & Hauling							
Loading, Unloading, Erecting, Etc.							
Maintenance							
Pumping							
Scaffolding							
Small Power Equipment/Tools							
Field Offices: Job Office							
Architect/Owner's Office							
Temporary Telephones							
Utilities							
Temporary Toilets							
Storage Areas & Sheds							
Temporary Utilities: Heat							
Light & Power							
Water							
PAGE TOTALS							

Figure 6–17
Project overhead summary sheet

153

DESCRIPTION							
Totals Brought Forward							
Winter Protection: Temp. Heat/Protection							
Snow Plowing							
Thawing Materials							
Temporary Roads							
Signs & Barricades: Site Sign							
Temporary Fences							
Temporary Stairs, Ladders & Floors							
Photographs							
Clean Up							
Dumpster							
Final Clean Up							
Punch List							
Permits: Building							
Misc.							
Insurance: Builders Risk							
Owner's Protective Liability							
Umbrella							
Unemployment Ins. & Social Security							
Taxes							
City Sales Tax							
State Sales Tax							
Bonds							
Performance							
Material & Equipment							
Main Office Expense							
Special Items							
TOTALS:							

Figure 6–17 (continued)

154

Overhead

Project overhead includes both certain field activities and home office costs. In the illustrated summary sheet example (Fig. 6–16), overhead is classified as General Requirements and is covered in the Division 1 line item. The overhead calculations illustrated in Figure 6–16 (at item 3, see percentages worked out below) covers the additional costs associated with the management of the home office of the business. Examples of costs associated with the home office would be office rent or real estate costs, vehicles, engineering support, clerical staff, top management salaries, and marketing, legal, and accounting fees. In a unit price estimate, home office overhead, also called project overhead, is factored into each of the four cost categories on the estimate summary: material, labor, equipment, and subcontractor. The percentages that are used vary based on the project type and the general contractor's cost of doing business. The percentages noted below are those suggested by the R.S. Means Co., based on national averages. The percentages that Means suggests include both overhead and profit.

Material and equipment costs are adjusted 10 percent to account for the costs of managing the purchase or rental, storage, and handling of the materials and equipment at the job site. Some risk is involved in getting the correct materials to the job site and in arranging for the proper equipment. The 10 percent overhead fee covers the costs associated with the home office support of this process and the profit associated with this work.

The overhead associated with labor is the greatest of any of the cost categories. The bare cost labor rate for each trade, which was shown in Figure 6–12, includes the worker's take home pay plus any fringe benefits, which would include vacation time and paid sick days. Added to these costs are the costs associated with workers' compensation insurance, which is figured at 19 percent, and federal and state unemployment costs, social security taxes, builders risk insurance, and public liability costs, which total 16.8 percent. These percentages are itemized on a national average basis in Figure 6–16. Fixed overhead, which covers the costs of maintaining the corporate office, as noted above, is 13 percent, and the profit billed out on labor is 10 percent. The average hourly rate for union workers, which started at a bare cost including fringes of $24.65, once overhead and profit have been added on increases 58.8 percent to a billable rate of $39.15.

The last cost category which must be adjusted for overhead and profit are the subcontractor bids. The subcontractor prices which are submitted to the general contractor already include the costs of managing the material and equipment purchases and rentals as well as the labor markups identified above, and include a fair profit for the subcontractor. Required sales tax should also be covered by the subcontractor within the bid price. The suggested markup for overhead and profit on the subcontractors' bids is 10 percent. This markup will cover the costs associated with organizing the bid packages, prequalifying subcontractors, reviewing bids, and managing the subcontractors' work in the field.

Profit

No contractor—no management of any business—is going to invest its time, energy, and absorb the risk inherent in a construction project for free. Profit is added at the point in the project when the contractor has quantified the work involved in the project and has priced the labor and equipment involved. Project overhead, also called general conditions, has covered the costs associated with managing the job in the field, and home office overhead covers the costs of supporting the project and the business at the home office. If no profit was added the business might "stay afloat," but it would not grow. The company would also have no financial tolerance for mistakes or unforseen conditions. (In Figure 6–16 profit was factored into the overhead that was computed; it could also be added into the contingency percentages at item 4, which, of course, also try to account for uncertainties over the course of the project.)

Good companies should and need to add a profit margin into each project. The amount of profit that is added is a factor of the type of project, its size, the amount of competition anticipated, the desire to get the job, and the extent of risk associated with the project. A good estimating process should identify the actual direct costs for the project, the indirect costs associated with project and home office overhead, and then allow top management the opportunity to factor in a correct profit percentage. The profit added on is based on the factors identified above and can range from 0 to 25 percent. If the company needs work, is faced with a lot of competition, or has a strong desire to move into a relationship with a new client, the profit charged may be small. A project of a high risk nature in a market in which a company sees little competition would allow a higher profit percentage. (Such a percentage was factored in to the example in Figure 6–16 under Adjustments, at item 5.)

Conclusion

This chapter has covered the steps involved in preparing a detailed, unit price estimate and completes this book's coverage of the estimating process. The detailed estimate can be prepared by the owner/management team to determine the fair cost of a project or by the contractor to bid on the work. Either estimator needs to work with a complete set of contract documents and needs to devote adequate time to the process. If a contractor's bid is accepted, the estimate used to prepare the bid will establish the contract price for the project.

An organized approach is the key to the preparation of an accurate estimate since many people and organizations will be involved in the process. The project needs to be broken down by bid packages along a divisional format which will allow subcontractors and specialists to focus on their areas of expertise. Accurate quantities and unit prices need to be established and totaled using material, labor, equipment, and subcontractor categories. Project overhead, usually covered under general requirements, can either be itemized or treated as a percentage. Sales tax and corporate overhead are generally treated as percentages and applied individually to the four cost categories identified above. Profit needs to be added on at the end; the amount of profit added into the project depends on the characteristics of the project. Profit can be added in with overhead as was done in the example in this chapter, or can be added as a separate line item at the end of the estimate. Though it was not discussed in this chapter, adjustments might still need to be made for location and time. These would be made at the end of the project.

Chapter Review Questions

1. Detailed estimates are made with very little information known about the project.
    __ T   __ F

2. Equipment costs are figured by totaling the cost of maintenance, fuel, and operator.
    __ T   __ F

3. Project overhead and home office overhead are the two major overhead components.
    __ T   __ F

4. The goal of the quantity takeoff process is to quantify every item of the project.
    __ T   __ F

5. The pricing of material unit costs is more difficult than the pricing of labor unit costs.
    __ T   __ F

6. Given a floor dimension of 24′ × 36′, how many sheets of plywood are required for subflooring? (A sheet of plywood measures 4′ × 8′.)
    a. 27
    b. 32
    c. 864
    d. 108

7. How many cubic feet are in a cubic yard?
    a. 9
    b. 18
    c. 27
    d. 36

8. What is the quantity in board feet of ten 2 × 8 ceiling joists? (Note: 1 board foot = 1″ × 12″ × 1′.)
    a. 160
    b. 16.6
    c. 13.3
    d. 12

9. Labor pricing requires the knowledge of:
   a. Activity duration
   b. Hourly wage
   c. Material unit price
   d. A and B only
   e. All of the above

10. The amount of profit charged on a project is a factor of:
    a. The extent of risk
    b. The amount of competition expected
    c. The expected duration of the project
    d. All of the above

Exercises

1. Describe the difference in approach to preparing a detailed estimate as a bidder versus as an owner or designer.

2. Describe the reasons why a site visit is important in preparing an accurate detailed estimate.

3. Identify the sources of unit price information for a contractor.

4. Obtain a copy of an estimate from a contractor. How was the estimate organized? Identify which prices were received from vendors and subcontractors. Examine how the contractor quantified the work that was done in-house. How were general conditions treated? How was home office overhead covered? How was profit built in?

Sources of Additional Information

Building Construction Cost Data. Kingston, MA: R. S. Means Co., Inc., Published annually.

Builders Estimator's Reference Book, 24th ed. Chicago: The Frank Walker Co., 1992.

Dagostino, Frank R. *Estimating in Building Construction*, 4th ed. Englewood Ciffs, NJ: Prentice Hall, 1993

Foster, Norman, Theodore J. Trauner, Jr., Rocco R. Yespe, and William M. Chapman. *Construction Estimates from Take-Off to Bid*, 3d ed. New York: McGraw-Hill, 1995.

Helton, Joseph E. *Simplified Estimating for Builders and Engineers*, 2d ed. Englewood Cliffs, NJ: Prentice Hall, 1992.

Means Estimating Handbook. Kingston, MA: R. S. Means Co., Inc., 1990.

Peurifoy, Robert L. *Estimating Construction Costs*, 3d ed. New York: McGraw-Hill, 1975.

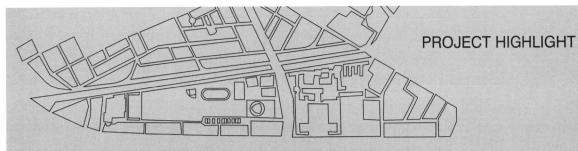

MIT Renovation of Building 16 and 56

SECTION TWO

Setting Up the Budget and Formatting the Estimate

Nancy E. Joyce

In any construction project, one of the first questions an owner seeks to answer is how much the project will cost. Because financing a major renovation is a significant undertaking, early accurate estimates are of prime importance. With the renovation of Buildings 16 and 56, the question of cost was interwoven with the question of how extensive the scope of work would be. Until the analysis of occupancy was complete, multiple approaches were considered, each of which carried its own costs.

Once occupancy was determined, a preliminary project budget was established. The construction cost made up the majority of the total project budget, but there were also many other costs the owner incurred. In a preliminary budgeting situation, the owner normally carries about 35 percent of the construction cost to cover these categories. These cost categories include consultant fees, equipment and furniture, telephone, security, audio/visual equipment, permits, regulatory fees, and hazardous waste. Initial costs incorporated into these categories were a combination of historical data and percentages. The total of all these categories comprised the project budget (see Figure A).

Project Cost Report - Detail

0100 Property Acquisition
0110	Property Cost
0120	Site Assessment
0130	Building Assessment
0140	Building Appraisal

0200 Site Improvements
0210	Hazardous waste disposal
0211	Asbestos abatement
0212	Clearing and Grubbing
0213	Building Demolition
0220	Water
0221	Sanitary
0222	Drainage
0223	Electric
0224	Gas
0225	Telephone
0226	Steam
0227	Cable TV
0228	Fire Alarm
0229	Automatic Temperature Control
0230	Traffic Control
0240	Grading
0241	Paving/Curbing
0242	Landscaping

0300 Construction
0310	Interior Demolition
0320	Base Building/Shell
0321	New Fit Out
0322	Renovated Fit Out
0330	Retainage
0331	Construction Contingency
0332	Escalation
0333	Phasing
0340	Construction Management Fee

0400 Special Construction
0410	Exploratory Construction
0411	Preliminary Construction
0412	Mockups
0413	Work by Owner
0420	Telephone/Data
0421	Security System
0422	Audio Visual Systems
0430	Construction Management

0500 FF&E
0510	Furniture Inventory
0511	Furniture Salvage Value
0512	Laboratory Furniture
0513	Classroom Furniture
0514	Systems Furniture
0515	Free Standing Furniture
0516	Specialty Furniture
0517	Custom Furniture
0518	Filing Systems
0519	Accessories
0530	Laboratory Equipment
0531	Classroom Equipment
0532	Office Equipment
0533	Kitchen
0534	Computers
0540	Graphics & Signage
0541	Bulletin Boards
0542	Coat Racks
0543	Artwork
0544	Window Treatment
0550	Temporary Move
0551	Permanent Move
0560	Benches
0561	Planters
0562	Flagpoles
0563	Fountains

0600 Architecture/Engineering
0610	Preliminary services
0611	Feasibility
0612	Programming
0620	Basic Services
0630	Additional Services
0640	Reimbursable Expenses
0650	Electrical
0651	Fire Protection
0652	Structural

0700 Consultants
0710	Surveyor
0711	Geotechnical
0712	Environmental
0713	Parking/transportation
0714	Air Quality
0715	Wind Tunnel
0716	Civil Engineer
0717	Asbestos
0720	Building Code
0721	Accessibility
0722	Public Processing
0723	Legal
0724	Schedule

Figure A
Categories used by the construction manager to formulate the project budget.

Project Cost Report - Detail

0730	Curtainwall		1220	Builders' Risk
0731	Building Systems		1230	Liability
0732	Furniture and Equipment		1240	Workmen's Compensation
0740	Graphics		1250	Bid Bond
0741	Acoustical		1260	Performance Bond
0742	Lighting			
0743	Energy		**1300**	**Municipal Assessment**
0744	Elevator		1310	Real Estate Taxes
0745	Low voltage systems		1320	Linkage
0746	Security			
0747	Audio visual		**1400**	**Financing**
0748	Kitchen		1410	Financing Fees
0749	Moving		1420	Interim financing
			1430	Permanent financing
			1440	Operating deficit reserve

0800 Quality Control

0810	Site representation		**1500**	**Administration**
0820	Utilities		1510	Program Management
0830	Geotechnical		1520	In-house Project Management
0840	Structural		1530	Other Administrative Costs
0850	Envelope			
0860	Mechanical		**1600**	**General Contingency**
0870	Elevator			

0900 Permits/Licenses/Fees

TOTAL PROJECT COSTS

- 0910 Environmental Processing fees
- 0920 Development impact fees
- 0930 Other public fees
- 0940 Building Permits
- 0950 Utility charges

1000 Project Close-out

- 1010 Final Testing
- 1020 Calibration Services
- 1030 Inspections
- 1040 Certificate of Occupancy fees

1100 Pre-opening

- 1110 Brochures - Promotional
- 1115 Scale Models
- 1120 Architectural Renderings
- 1125 Audio/Visual Presentations
- 1130 Broker Promotion
- 1135 Promo, Entertain & Travel
- 1140 Marketing Office
- 1145 Media Advertising
- 1150 Project Promo Signs
- 1155 Events

1200 Insurance and Bonds

- 1210 Title Insurance

At the completion of schematic design, the project estimate was established. This consisted of the construction estimate plus the other owner costs. To develop the construction estimate, the owner elected to retain two cost consultants in the design phase of the project. The construction manager was very familiar with the local marketplace and had a solid data base of similar jobs that had been done recently. They did not, however, have extensive experience with the architect. MIT asked the architect to retain a cost consultant familiar with their work. By doing this, the owner was insuring that assumptions being made early in the design period by the cost consultants would correspond to the level of design detailed by the architect during later phases of the design. The use of two independent estimators provided the owner with an appropriate check and balance approach.

Before the estimators put together their estimates, the two estimating teams met to discuss how to formulate the categories of costs so that their independent estimates could later be compared on a line by line basis. Because this was an early estimate, the estimators chose a building system format. Utilization of this breakdown facilitated the use of assembly costs and square foot costs. Also, by using a common breakdown of costs, the two teams were able to more easily reconcile their numbers. The reconciliation was accomplished through a series of meetings between the estimators after they completed their individual estimates. At these meetings the two parties focused on areas where the differences between line items was greater than 10 percent. Both discussed the assumptions made while preparing the estimate. The project team leaders were also present at these meetings. The architect confirmed the design intent in areas of differences, the construction manager discussed the construction and cost implications, the owner confirmed Institute priorities and standards, and the estimators made adjustments as appropriate.

After all the assumptions were confirmed and the estimators revised their estimates, the project team reconciled the final numbers by discussing the differences remaining and deciding how to carry the particular line item, usually by weighing factors on one side or the other. In this way, the entire team became familiar with the scope of the project, the intent of the designers was laid out, and the final estimate gained the wisdom and experience of all members.

As part of the schematic design estimate, a series of alternatives were developed. Some were deduct alternatives and some were add alternatives. In addition, the estimators were asked to submit a list of cost reduction recommendations. These, along with the alternatives, were compiled on a worksheet so that the team could discuss and analyze the merits of each. During design development many changes occurred which affected the cost of the project. User requests for specialized spaces, changes in program, requirements by local authorities, and aesthetic considerations all put pressure on the budget. Acceptance of cost reduction recommendations helped to offset these and maintain the budget.

There were also some areas of the project that were examined separately because of the cost and schedule implications associated with them. One example of that is the curtainwall system. This was original to the building but appeared to be in good condition. An extensive study was done with a curtainwall consultant and it was determined that with repair and periodic maintenance the curtainwall could be retained. Retaining the curtainwall minimized the disruption to the surrounding campus activities. Without removing the curtainwall, most construction could be confined to the interior of the buildings. Because of the long lead time needed to design and purchase a new curtainwall as well as the significant cost involved, it was important to get more specific information about this early in the project (see Figure B).

At the completion of design development, the owner again retained two cost estimators. Because the construction manager was primarily submitting packages to subcontractors for pricing, the format of the estimate was changed to reflect the traditional division of work packages. This was done through the use of the 16 CSI divisions. The use of this format would also simplify comparison of numbers submitted at bid time. The cost consultant retained by the architect continued to price the job primarily utilizing their data base. Utilization of these two different methods of estimating served to confirm each number. The process of reconciliation was similar at the end of design development as it had been for schematic design. However, more information was available about the project and the scope of work was fairly well defined. Discussions revolved more on particular unit prices carried and on the quantities assumed.

Because the reconciled design development estimate confirmed the schematic design estimate, the design team was directed by the owner to proceed with the construction documents stage. The focus of the project team shifted from examining the pricing of individual cost reduction items to maintaining the budget through the detailing of construction documents. Cost reduction recommendations continued to be examined during construction documents but were accepted on merit alone and included as part of the construction documents without separate pricing.

At the end of design development, the owner accepted the scope of work and level of quality shown in the drawings. The team worked to maximize value throughout the various phases of the design process and felt confident that MIT was getting the right product at the right price. When the documents went out on the street, the marketplace would be the final determinant that our beliefs were well placed.

Cost Summary - Building 16

Cost	Repair	New
First Cost		
Mockups	15,000	
Remove existing		45,000
Remove lead paint, Full prime, 1 finish coat	366,400	
Wet seal	111,450	
Weatherstrip	37,000	
Aluminum cap	75,000	
New Curtainwall		1,234,800
Subtotal	**604,850**	**1,279,800**
General Conditions and Fee (15%)	90,728	191,970
Subtotal	**695,578**	**1,471,770**
Soft Costs (35% and 20%)	243,452	294,354
Total Project Cost	**939,030**	**1,766,124**
Alternate		
Patch/paint	(345,975)	
Maintenance over 30 years (in current dollars)		
Repaint - 15 years	316,400	
Reseal - 15 years	111,450	
Weatherstrip - 15 years	37,000	
Markups - 15 years - 25%	116,213	
Spot seal - 5 years @ $50,000	200,000	250,000
Total Maintenance	**781,063**	**250,000**
Energy Savings over 30 years (in current dollars)		
$0.60/sf x 29500 sf x 30 years (excl. 8th floor)		(531,000)
Life Cycle over 30 years		
Net Present Value discounted at 8% annually	**1,423,453**	**1,634,827**

Figure B
Curtainwall cost summary for Building 16. (Life cycle costing examining the impact of keeping the curtainwall over a 30 year period)

Cost Summary - Building 56

Cost	Repair	New
First Cost		
Mockups	15,000	
Remove existing		68,000
Remove lead paint, Full prime, 1 finish coat	553,600	
Wet seal	168,550	
Weatherstrip	N/A	
Aluminum cap	80,000	
New Curtainwall		1,910,000
Subtotal	**817,150**	**1,978,000**
General Conditions and Fee (15%)	122,573	296,700
Subtotal	**939,723**	**2,274,700**
Soft Costs (35% and 20%)	328,903	454,940
Total Project Cost	**1,268,625**	**2,729,640**
Alternate		
Patch/paint	(523,425)	
Maintenance over 30 years (in current dollars)		
Repaint - 15 years	503,000	
Reseal - 15 years	168,550	
Insulating glass unit replacement - 15 years	210,000	
Markups - 15 years - 25%	220,388	
Spot seal - 5 years @$70,000	280,000	350,000
Total Maintenance	**1,381,938**	**350,000**
Energy Savings over 30 years (in current dollars)		
Not applicable		N/A
Life Cycle cost over 30 years		
Net Present Value discounted at 8% annually	**2,123,681**	**2,951,962**

Figure B (cont.)
Curtainwall cost summary for Building 56

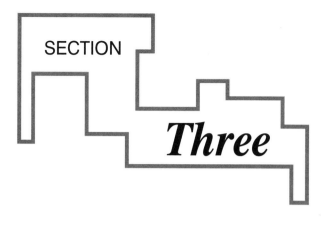

Three

Scheduling

This section of the book addresses the subject of scheduling. It looks at the reasons why schedules are produced and the types of schedules that are used. Schedules are produced throughout the life of a project by many of the project team members. Scheduling, like estimating, is a basic project management skill. Their two end products should be developed concurrently. In fact, accurate estimates need to use the information that is developed within the schedule.

Chapter 7 will provide a basic overview of the scheduling process and then will look at when schedules are developed and the types of schedules that are used. Chapter 8 and 9 will explain the details of network schedules, also called CPM schedules. Chapter 9 will cover network calculations, including how activity durations are calculated.

CHAPTER

7

Scheduling Fundamentals

Introduction

Schedule Definition

Scheduling is the process of listing a number of duties or events in the sequence that they will occur. It is a timetable, and it formulates the activities that must be accomplished to reach a certain goal or objective.

Schedules are essential to the successful execution of any project. They are not unique to construction, as they are used in many industries—business, manufacturing, publishing, and so on. Anytime that people, equipment, materials, and organizations are brought together and directed towards a common goal, a schedule will be used.

History of Scheduling

The formalized use of schedules has been in existence since the 1950s. The E. I. du Pont de Nemours Company in 1956, using a UNIVAC computer, developed a Critical Path Method (CPM) schedule for a $10 million chemical plant in Louisville, Kentucky. At about the same time, the U.S. Navy used a Performance Evaluation and Review Technique (PERT) network schedule to manage the development of the Polaris missile. PERT and CPM, both network based scheduling systems, were also used throughout the 1960s by the Corps of Engineers, NASA, the Atomic Energy Commission, RCA, General Electric, the Apollo program, the Veterans Administration, and the GSA.

During the formative days of scheduling, formalized schedules were utilized on only the largest of projects and were developed by people sophisticated in engineering principles and early computer technology. These schedules were run on large, mainframe computers, which by today's standards were difficult to operate. Today, because of the widescale use of personal computers with their simplified hardware and software, and with more sophisticated owners, designers, and construction professionals, network schedules are much more widely used. Not only are they commonly utilized in the planning stage, but they have also become a basic part of the control system of most construction projects. Network based schedules are considered acceptable evidence in court when arguments occur over project completion dates, delivery dates, or the formal coordination of project participants. Conflicts can occur over many issues during the course of a construction project. Design changes, poor weather conditions, labor actions, mistiming of deliveries—all these are a common basis of conflicts. If the parties involved are unable to work out their differences through negotiation or arbitration, the alternative is to use the court system. At that level, judgements that involve time and project coordination would be based on the network schedule used.

When to Schedule

Schedules establish the start, duration, and completion date of a project or a task. They let people and organizations know in advance when to expect a certain action to take place. Every contractor and subcontractor must perform their work profitably. To maintain a profitable business, these contractors will have many jobs going on that all have to be collectively organized. When a particular job is going to begin is vital information and so is the expected completion date. With this information the subcontractor can know whether or not the work of a particular job can be accomplished in the context of all the other work scheduled. In addition, the firm needs to know precisely when other work that it is dependent upon will be complete, since this part of the project cannot be finished until the work of these other contractors is complete. Knowing precisely when an activity is going to take place also has substantial cost implications. For instance, a large crane can rent for as much as $2,500 per week (see Fig. 7–1), so if the duration of a project is not figured closely a contractor can quickly consume in rental charges any profit it would hope to get from a job. A contractor's overhead is also dependent on how long a project is expected to take. Examples of overhead costs dependent on project duration are rental for site fencing, salary for the job superintendent, and maintenance of a field office.

Scheduled start dates determine when goods and services need to be brought to the job site, when a work force needs to be mobilized, and when equipment rentals begin. This date is critical to the accurate pricing of the project. A delay in the start of the project could significantly affect the cost of material, and the rate at which labor can be bought. Both of these are significant parts of a contractor's estimate and are priced with an understanding of when they will be bought. Also, materials brought to the job site early can be lost, stolen, or vandalized. In the case of a job site where the set up and storage space is restricted, such as a downtown site (see Fig. 7–2), material deliveries must be closely coordinated for immediate usage. There are times when contractors want key materials delivered early to eliminate the risk of the project being held up because of delays in getting the material, but even then they must accurately budget storage costs and therefore need to know how long it will be before the material will be installed on the project.

Figure 7–1
A crawler crane can rent for as
much as $2,500 per week.
Courtesy of Rotondo Precast
Photo by Don Farrell

Preconstruction

Scheduling is an important activity during the preconstruction stage of a project. Owners need to know upfront if the project can be completed on time. Owners, to secure project financing, must establish firm commitments with the end user of the finished project. In the case of a highway, this would be the public; in the case of a strip mall development, a private retail tenant. A retail operation which misses the Christmas shopping season will experience heavy losses. The world's athletes were seriously affected when Montreal's Olympic Stadium was not finished in time for the 1976 Olympics. (See sidebar telling this story.) The Public Broadcasting Service network aired a series called *Skyscraper* in 1990 in which construction delays prevented the on time occupancy of a skyscraper by a key tenant. William Zechendorf, Jr., the developer, had signed a contract with this tenant early in the project. When the project was delayed, the space was not delivered per contract, obligating Zechendorf to damages (for lost rent plus damages) at the rate of $300,000+ per week. This obliga-

Figure 7–2
Job sites can be busy and congested, requiring the exact scheduling of all deliveries.
Courtesy of Walsh Brothers, Inc.
Photo by Don Farrell

tion could have been avoided by a more realistic projection of the date that tenant occupancy could have occurred. The only way to accurately predict whether or not the required completion dates are meetable is by the use of a schedule. This schedule must accurately identify all the tasks required to be completed on a project, determine how long each will take, and place them in the correct, logical order.

Even with the completion date of a project identified, successful project managers should still look to optimize the project. Through closer examination of the initial schedule, efficiencies in time and money can be identified. Oftentimes key subcontractors are not available at certain times. It may make sense for the project manager to reschedule their work to a more opportune time. Some work forces may be overcommitted during a certain period. In this case, because of resource constraints it may be necessary to move certain activities to a different time. If that is not done either the project will be delayed or additional hiring costs will be incurred as the necessary people are brought in. It may make sense to combine certain operations during a particular week. Key project resources such as a crane, backhoe, or management team may

Sidebar ▄▄

Montreal Olympic Stadium

On July 17, 1976, after the boycotters had left and the politicians had made their last wearisome demands, the Games of the XXI Olympiad opened under bright skies in Montreal. Only a small fraction of the vast, worldwide television audience who watched the colorful and carefully choreographed opening ceremonies could have guessed that, only a short time before, the gleaming Olympic site had been an absolute shambles. Eighty cranes and 2700 workers, in a construction saga with few equals for anguish, had been hurled against the fast closing Olympic demand. In chaotic, round-the-clock shifts, they brought the giant stadium to some semblance of readiness—a stadium that had cost almost $800 million to build, a stadium that had become the center of so much controversy and the symbol of so much extravagance.

> Barclay F. Gordon
> *Olympic Architecture: Building for the Summer Games* (New York: John Wiley and Sons, 1983), p. 136.

Montreal's Olympic Stadium is an excellent case study of how a project with a defined need, established budget, and adequate timeframe can still fail. The Olympic project entered the planning stage about 1969. The package consisted of a velodrome, stadium-mast-pools complex, parking areas, a generating plant, outdoor facilities, and landscaping. A four-volume report prepared by The Commission of Inquiry (established in 1977 by the government of Quebec upon the recommendation of the prime minister) documents quite clearly why this project experienced such substantial cost overruns.

The first documented price of $120.5 million had been submitted by the mayor of Montreal, Mr. Jean Drapeau, to the Canadian Olympic Association in 1969. As of 1976, at about the time the XXI Olympiad opened, the documented cost of the project had grown to $1.333 billion, with the facilities even at this point still incomplete. The cost to complete this project, according to Quebec's minister of finance, was an additional $137 million.

How and why did this project fail?

The Commission of Inquiry cited many reasons, but the most important was a complete lack of construction project management. The project from its conception in about 1970 until two years before the games' opening, when a formal project management team was assigned, had been managed by the mayor of Montreal, Mr. Jean Drapeau. The mayor, working very closely with the principle architect, Mr. Roger Taillibert of Paris, France, abandoned the idea of a modest facility as originally proposed and estimated. The design that was pursued was dictated by aesthetics and was one of grandeur. Only late in the project was a formalized master plan and budget established. This state of affairs had provided the mayor and the architect the freedom to design and build the Olympic complex with no monetary constraints until late in the project.

The lack of a formal project management system is evident in the total lack of control exhibited in the project. Aesthetics and grandeur were the key criteria to both the architect and the mayor; no other decision makers had been involved until later in the project. One can see how the project cost and schedule were sacrificed. If a budget and schedule had been set, then design decisions could have been measured against these restrictions. As it was, a very complicated precast concrete system was chosen. Duncan Robb, a consulting engineer later involved in the project, deemed the Olympic installations to be among the most complicated buildings in the world. Even the Canadian National Tower in Toronto, considered to be one of the most advanced structures in the world from a technical standpoint, is really quite simple when compared with the Olympic velodrome.

The selection of an architect from another country also complicated the process. The role an architect plays in the design and construction process and the technologies readily available in the two countries differed, creating problems of coordination. The roles of the engineering consultants became confused, causing redundant engineering and poor design quality. At one time it was found that two consulting engineers were hired at the same time to do the same design and both were paid. Communication between the construction professionals and the designer was also poor, particularly with respect to constructability. The precast concrete system used was unfamiliar to the contractors in the area. Given the extremely cold weather conditions in Montreal it became a costly installation. In addition to all metric dimensions (Canada at that time had yet to adopt the metric system), the architect insisted that a European epoxy-gluing and post-tensioning system be used, which was new to North American builders.

As Gordon puts it,

> Threading the miles of cable through the subsections had its own complications. At several critical locations, cable accumulated in awkward amounts. . . . None of this work was made easier, of course, by the near-Siberian cold. In spite of such precautions as were possible under the circumstances, ice and surplus epoxy hardened in many post-tensioning channels. Before work could resume, these channels had to be painstakingly cleared by men who were themselves in danger of frostbite. (*Olympic Architecture*, pp. 143-44)

Even though a detailed network base schedule was specified as early as 1970, no schedule was used during the entire preconstruction stage. The earliest CPM schedule found was prepared by the building contractor on the velodrome. The first overall CPM schedule was produced by the coordinating project manager, who was hired in the fall of 1974, less than two years before the Summer Olympics was to begin. The lack of adequate scheduling is reflected in the drastic difference in the planned time to study and construct the specific buildings versus what was actually spent. For instance on the velodrome 6 months were planned to study and 13 months to build, whereas 12 months were actually used to study and 35 months to build. Because of these delays the project involved three winters instead of two, and showed further escalated prices due to inflation.

A solid management team was most missed in general management of the project. The project suffered from galloping inflation, a saturated construction market, strikes, work stoppages, corruption, and fraud. The estimating process was inadequate for this type of project. The project was a nonstandard type, with the unit prices used being appropriate for more conventional projects.

In summary, the project failed. The 1976 Olympics were conducted in facilities that were not complete. If an adequate project management team had been hired, a formal system of controls would have been established. The scope and scale of the project would have been balanced by the time required for completion and the available funds. The design of the project would have been subjected to a constructability analysis questioning the complicated precast concrete design and metric dimensioning. Agreements with labor could have been arranged, inflation better projected, and corruption and fraud combatted.

be on site already for another activity and may be underutilized. By scheduling these operations concurrently, additional efficiencies could be gained. The initial schedule is but a tool for the project team to begin to make intelligent management decisions.

The scheduling process during the preconstruction stage should be viewed as an opportunity to design and build the project "on paper" prior to the actual construction. This provides an opportunity for all project parties to visualize the process and to make all the necessary provisions to properly coordinate the entire process. This is the time that the project team may decide to order key long-lead purchase items such as structural steel, elevators, or compressors. The project team may determine at this time that it will be necessary to begin construction before the project is totally designed (fast-tracking, as described earlier in this book) to shorten the delivery time for the project.

In summary, scheduling in the preconstruction stage is important to provide the owner with the necessary information to properly plan and coordinate the entire design and construction process. Knowing the exact dates when all the key events are going to occur is critical to the overall success of the project. Most construction projects must be worked around public commerce and existing operations and involve many design professionals, regulatory agencies, financial institutions, and ultimately the end users, all of whom are keenly interested in when they will be involved in the project and for how long. These answers can only be provided through the use of a schedule.

Construction

Project schedules are not only useful during the preconstruction stage, they are essential to the successful coordination of the day-to-day activities of a project. Material deliveries and the utilization of equipment and people are all orchestrated through the schedule. As a project progresses, delays inevitably occur. The project manager's job is to effectively deal with these delays and to anticipate them as much as is humanly possible. If problems didn't occur on a project, there would be little need for the services of a project manager. Delays are inevitable. It is the intelligent response to bad weather, equipment failures, strikes, design errors, or omissions that separate the well-managed project from the disaster.

During construction at the job site, a frequent use of the CPM (Critical Path Method) schedule is to record the actual activities at the site on a day-to-day basis. The schedule often is placed on the wall of the job site trailer where it is clearly visible to all the trades. The project manager can graphically record progress (see Fig. 7–3). This is often done with different color markers and symbols. This practice cannot only record issues for the day, it can help anticipate problems that may occur in the future.

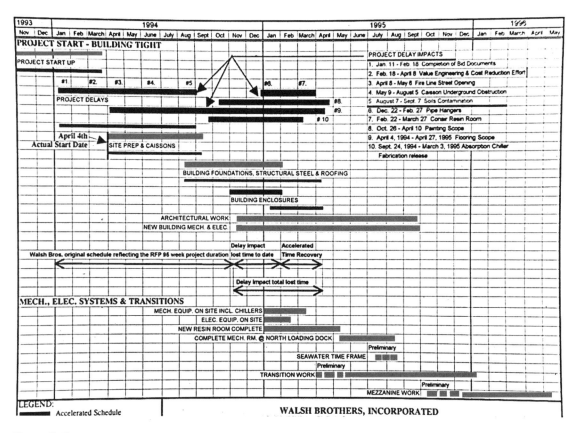

Figure 7–3
As a construction project proceeds, key project activities should be indicated on the construction schedule.
Courtesy of Walsh Brothers, Inc.
Photo by Don Farrell

This information may also prove essential to the successful negotiation of a future change order or delay claim. Just remember that tomorrow's lawsuit may be occurring today. Written documentation in a court of law has much more weight than someone's memory (see sidebar, The Use of Construction Schedules in Claims and Litigation).

Postconstruction

As a project nears completion the ultimate user of the facility becomes more involved in the construction process. In many projects the owner begins to occupy the facility while construction is still occurring. This is called **partial occupancy,** and if this is to occur it must be closely scheduled, requiring weekly meetings to coordinate the construction work with the tenant improvement work necessary to allow occupancy.

Most projects require testing and acceptance of equipment, the training of the people that will ultimately use and maintain the equipment on the project, and the correction of deficiencies, also called a punchlist. All of these need to be coordinated and controlled to occur smoothly.

An important control function is for the project team to "close the books" on the project. If people and the companies involved are to learn from the project experience, they need to record the actual events of the project. Events not planned for (e.g., coordination difficulties, production rate differences, delivery delays) should all be recorded. This actual data should be stored so that in the future better planning and scheduling can occur. Record keeping is also important in the event that any claims or disputes occur in the future.

Scheduling Methods

A schedule is a tool; it can be used to manage, coordinate, control, and report. Depending on the sophistication of the user, the schedule can take different forms. For instance, the owner/developer of a $500 million high-rise does not need to get involved in the coordination of the drywall finish work, but would certainly be interested in when the project is due for completion, as well as how the progress of the project is proceeding as compared to overall schedule and budget. This type of information can be best provided through the use of a bar chart. A bar chart can be developed quickly and inexpensively and is simple enough that its reader does not need any special training.

The field superintendent or the construction manager, on the other hand, is concerned about the delivery of key materials as well as the coordination of the many subcontractors on the project. The job involves ensuring that the work proceeds as planned and that no one particular event disrupts the flow of the project. To

Sidebar ▬▬▬▬▬▬▬▬▬▬▬▬▬▬▬▬▬▬▬▬▬▬▬▬▬▬

The Use of Construction Schedules in Claims and Litigation

In addition to being useful tools for project execution, construction schedules can also serve as weapons in a war of construction claims. Contractors whose work has been delayed or disrupted will have a chance of collecting additional compensation if, and only if, they can show the causes and effects of delay through clear, graphic, schedule-based documentation.

Specialized claims consultants or other scheduling experts are often engaged to help the parties to a construction dispute analyze, quantify, and present claims and defenses. Though their methodology varies, these experts usually base their analysis on a comparison of three types of construction schedules:

1. An as-planned schedule, showing how the contractor intended to construct the project within the originally established contract time.

2. An as-built schedule, showing what actually happened in the field.

3. Some kind of adjusted or impacted schedule (of which there are many varieties), showing how various delays and disruptions affected the as-planned schedule and/or contributed to the as-built schedule.

Different kinds of delay have different legal and financial impacts. An *excusable* delay, not the fault of the owner or contractor, will entitle the contractor to an extension of time but no additional compensation. A *non-excusable* delay, due to the fault of the contractor, will result in no extension of time, and may entitle the owner to collect actual or liquidated damages from the contractor. A *compensable* delay, due to the fault of the owner, will entitle the contractor to both an extension of time and additional compensation.

If there are concurrent delays of different kinds, they will often cancel each other out, resulting in no compensation or damages to either the owner or the contractor. However, there may be circumstances that make it possible to apportion the effects of concurrent delays, or to give one type of delay precedence over the other. The strength of the contractor's case or an owner's defense will often turn on how skillfully the scheduling expert can demonstrate the causes and impacts of the various kinds of delays.

For instance, an owner-caused delay may have occurred during a period of time when the project would have been delayed in any event by the contractor's failure to order materials in a timely manner. Were they both on the critical path? Were they independent, or was one a reaction to the other? These questions can only be answered by a detailed schedule analysis and a careful review of the project records.

Schedules that are coordinated with manpower and cash flow information can help the contractor to document costs that may not be immediately apparent, such as the cost of "acceleration" (e.g., overtime costs incurred to meet an originally established schedule, where the owner refuses to grant an extension of time to which the contractor is entitled), or the cost of "loss of productivity" (e.g., additional effort required when two trades are forced to operate in the same space concurrently rather than sequentially). A contractor may allege that owner-caused delays are causing acceleration or loss of efficiency, but without clear and complete schedule documentation the contractor will stand little chance of collecting such costs in a negotiation, arbitration, or lawsuit.

Christopher L. Noble
Construction Attorney
Hill & Barlow
Boston, Massachusetts

manage this kind of process it is essential that a schedule be prepared that accurately reflects the detail of the actual project. A CPM schedule that is network based provides the necessary detail the superintendent needs. This schedule is expensive to produce and requires a fairly high degree of technical competence, but it provides the degree of information necessary to adequately control the project.

One of the drawbacks to the use of network based schedules is that they require technical sophistication, as mentioned above. To address this problem, scheduling methods such as matrix schedules and time-scaled bar charts and others have been developed. These scheduling methods combine the graphical benefits of a bar chart with the technical detail of a network schedule. The major goal of these scheduling methods is communication. Most people, technical as well as nontechnical, are able to read these schedules.

The fact that project people at various levels of a company require schedule information of varied levels of detail and precision has created these different methods of scheduling. Many scheduling methods exist, but the most common methods—and the ones that will be discussed in detail in this text—are the bar chart; linear balance, time-scaled bar, and matrix; and network based schedule.

Bar Chart Schedules

The bar chart, also called a Gantt chart, is graphically the most simple of the scheduling methods. It is understood by most project people and can be produced quicker than any of the other scheduling methods. It is frequently used in the planning stage of a project by owners, designers, and construction professionals to quickly examine the overall timing on a project. In its most simple form, an overall project may be

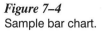

Figure 7–4
Sample bar chart.

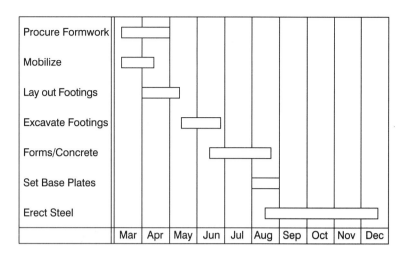

broken down into three bars reflecting design, bid and award, and construction. In the bar chart shown in Figure 7–4 one can easily learn several important facts about the project, including:

1. The planned overall length of the project
2. The planned duration of each project component (e.g., mobilize, lay out footings, excavate)
3. The calendar start and finish dates for each project phase

Bar charts can also be used to report information to people who are concerned about a project, but who may not be involved in the day-to-day management. Bar charts can provide a quick, visual overview of a project, but they tend to neglect the management detail necessary to make complicated coordination decisions. Bar charts can be color coded or time scaled; they work nicely as a tool to compare actual progress to planned. Bar charts are universally accepted, with the reader needing little, if any, specialized training. Bar charts are best used in conjunction with network based scheduling methods. The network based method is used by the scheduler and other project management people to lay out in detail the workings of the project, with the bar chart then used to communicate the results. It is important to remember that a bar chart does not communicate the interrelationships between project activities (see Fig. 7–5).

Because of the inherent graphic limitations of a bar chart, it cannot define individual activity dependencies. For instance, in Figure 7–5 it is not clear from the bar chart whether the Excavation of the footings is dependent on the completion of the Layout of the footings. Common sense and personal experience tells us that it is dependent, but the bar chart does not by definition define this dependency. Therefore, because these dependencies are not considered, bar charts cannot be used to calculate specific project activity start dates, completion dates, and available float (extra time available).

Figure 7–5
A problem with bar charts.

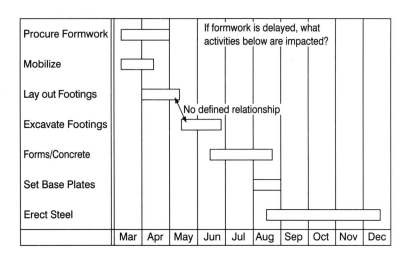

In summary, bar charts are excellent communicators of time-related project information. They are quick and easy to develop and are understood by most people. Their major limitation is that interdependencies between activities cannot be shown. Because that information is not provided, complicated management decisions should be made utilizing other, more thorough scheduling methods.

Linear Balance, Time-Scaled Bar, and Matrix Schedules

The linear balance schedule (see Fig. 7–6) incorporates some of the visual characteristics of a bar chart schedule with some consideration for the interrelationships between project activities. A linear balance schedule is used to efficiently plan out repetitive operations. Repetitive operations are visually plotted (the slope of each line indicates the activity's work rate), balancing work so that the rate of work of each operation is in balance, avoiding work delays or stoppages.

Figure 7–6
Linear balance.

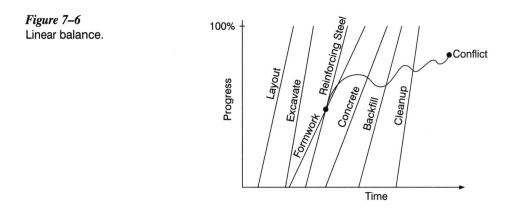

The time-scaled bar chart (see Fig. 7–7) is an expansion on the simple bar chart shown earlier. The X axis can be expanded or contracted as required. This bar chart is often logic based, identifying activity relationships, the project's critical path, and is sorted by activity codes. When used in this manner, the time-scaled bar chart serves as a way to present complicated, network generated scheduling information in a clear, concise manner. Notes, dates, or other pertinent information are often shown on this schedule.

Matrix schedules (see Fig. 7–8 and 7–9) are typically used where work is accomplished in a repetitive manner, such as on a high-rise office building. A quick review of this schedule provides the management team with an overall view of the project with consideration of the interdependencies between listed activities. Subcontractors or project people responsible for a specific task need only look at their specific

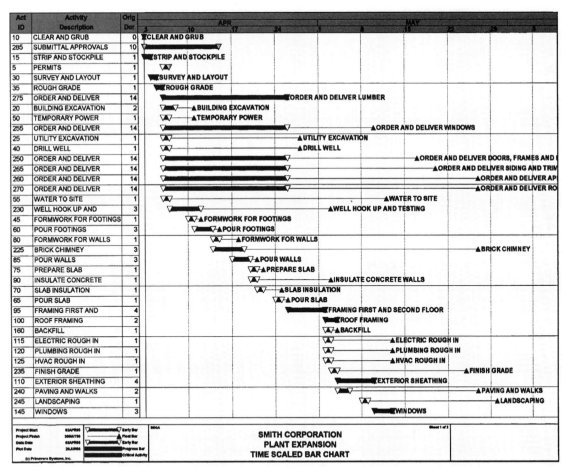

Figure 7–7
Typical time-scaled bar chart.

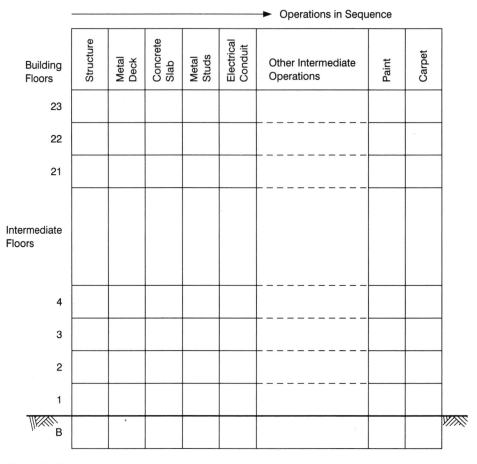

Figure 7–8
Example matrix schedule.

Figure 7–9
Typical cell matrix schedule.

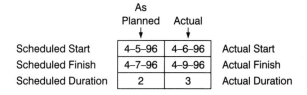

	As Planned	Actual	
Scheduled Start	4–5–96	4–6–96	Actual Start
Scheduled Finish	4–7–96	4–9–96	Actual Finish
Scheduled Duration	2	3	Actual Duration

responsibilities and see what precedes and succeeds their work. In Figure 7–8 the metal studs follow the concrete slab and precede the electrical conduit. The crew would also see that the work on the third floor precedes work on the fourth. By examining the schedule closer the crew can also see on what date the work is scheduled to start and finish (see Fig. 7–9). This figure shows the typical contents of a cell in such a matrix schedule. The planned start, finish, and duration of each activity is defined before the job begins; the actuals are plugged in as work proceeds.

A matrix schedule serves as a good tool to control the field activities of a project, as it can be posted at the field office and updated as the work proceeds. Superintendents can easily color in the boxes as the work is accomplished to provide a more visual picture of the job's progress. Supervisors responsible for a specific trade are able to critique their own progress in relation to related work.

Linear balance, time-scaled bar, and matrix schedules would generally not consider all project activities. They are best used as coordination schedules to communicate with field or office personnel. They can be used for presentation purposes, since they present information in a way that can be easily understood by nontechnical people. All of them present information in a manner that allows self-correction. Most project people can find a specific task on the schedule, see what the due date is, and what activities precede and succeed the activity.

Network Schedules

The workhorse of construction schedules is the network schedule. It is best prepared by a team of people who have complete knowledge of all aspects of the project. A completed network schedule means that all the work to be performed on the project has been defined and organized. In network scheduling each item of work is called an activity. These activities are each given a duration (how long the activity will take in hours, days, or weeks), and they are connected in what are called network diagrams. The completed network then defines all activity interrelationships and durations and considers what resources are available, as well as all assumptions about how the project will be pursued. A network schedule can be viewed as a road map that, if followed, will bring the project to its desired destination.

A project team which diligently prepares a network (commonly called a CPM) schedule has readied itself for the effective management of the project. Preparing a network schedule is like preparing an estimate. It forces a thorough review of all the contract documents as well as communication with the leaders of the forces that will be involved in the project. Questions such as what work can be scheduled concurrently or what task precedes the placing of the floor tile are all answered by the network schedule.

As can be seen in Figure 7–10, network schedules can take two forms, activity on arrow notation or activity on node notation, also called precedence notation. In activity on arrow notation the work or activity is shown on the arrows, which are connected by nodes. In precedent notation the work occurs on the nodes, and these are connected by arrows.

Both forms of network schedules are used in industry, with the end results identical. In the network scheduling packages Primavera and Timeline, activities are displayed as bars, with the beginning and end of each activity bar connected to the respective predecessors and successors. The following chapters will primarily focus on the theory of network scheduling, with exposure to both arrow and precedence notation.

The preparation of a network schedule is like building the actual project on paper. The schedule preparer must identify all the necessary tasks and then logically arrange them as the work will be accomplished in the field. The process, if

Figure 7–10
Network schedules.

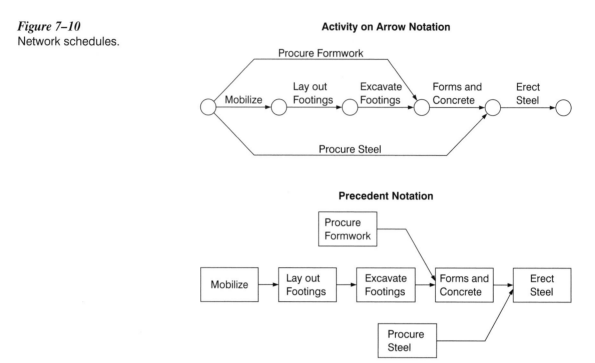

Activity on Arrow Notation

Precedent Notation

approached correctly, does and should force controversy, as there is usually more than one way to build a project. Discussions will occur both about the order of the activities and about the duration for a given task. These discussions are good as they force the project team to consider other options. This process is an important part of value engineering, allowing the project team to correctly consider project time during the planning stage of a project. I think all would agree that it is better to argue about the planned approach to the project in the office, on paper, before the project begins, than during the actual construction (see sidebar, Value Engineering, in Chapter 2).

When a network schedule is compared to the previously discussed matrix and bar chart schedules many differences appear. The network schedule is clearly the largest undertaking. The thorough preparation of a network schedule for a large commercial project—like a detailed estimate—can take several weeks and consume tremendous resources. Fairly detailed project information must be available, and to provide for the opportunity for updates and revisions, computer hardware and software must be owned or purchased. The preparation and interpretation of a network schedule requires technical training (see Fig. 7–11), which is why matrix and bar chart schedules are derived from the network schedules to communicate schedule information to both field personnel and nontechnical people such as the public and financial backers.

Figure 7–11
Preparation of a network schedule requires both technical and computer training.
Photo by Don Farrell

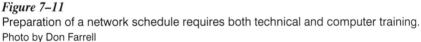

When network schedules first appeared in the construction industry they were viewed as an optional resource available to the contractor who wanted to invest in the technology. This was before the proliferation of the personal computer and its low cost computation power, which provides the contractor the ability to produce a schedule, update it, and produce reports all for little cost.

Owners, designers, construction managers, and other interested parties have come to understand the benefits provided by networks. Most project managers are now educated in scheduling theory and know how to use computers to produce network schedules. That is why many, if not most, major construction projects require that a network schedule be submitted before any construction can begin. This schedule may be developed independently by the contractor if the project is bid, or in the case of a construction management or negotiated approach the owner, designer, and construction professional may develop the schedule jointly. In any case, before the work begins, this schedule is complete and on hand to monitor the progress of the work. The impact of any delays, changes, or natural disasters can now all be compared to the baseline schedule. This provides the opportunity to prepare thorough reviews of the project status before making any adjustments to the project plan.

Conclusion

It should be clear from this chapter's discussion that the schedule is both a powerful management and communication tool. Because of the increased complexity in construction techniques and materials, as well as the diverse labor issues and the pressures of budgets, the use of this tool has increasingly become the standard control method. Without the use of a schedule it is difficult, if not impossible, to coordinate the diverse activities found in a construction project. An effectively managed project must closely coordinate the activities of the owner, designer, construction manager, and all the people who come together at the job site. Questions such as the owner wanting to know when the move can be scheduled, or the designer wanting to know if a certain lobby detail can be changed without delaying completion, or the electrical subcontractor wishing to know if they can substitute for a long-lead fixture can only be studied and answered by the use of schedules. In a construction project where time truly equals money the management of time is critical, and the best way to manage time is through scheduling.

Chapter Review Questions

1. Networks are continuous.

    __ T   __ F

2. Project scheduling techniques such as CPM are unique to the construction industry.

    __ T   __ F

3. Network schedules provide a more readable format than bar chart schedules.

    __ T   __ F

4. Matrix schedules are used to provide easy readability while continuing to show project logic.

    __ T   __ F

5. The production of an effective project schedule is a team effort.

    __ T   __ F

6. A construction project network shows:
   a. The order in which construction tasks must be completed
   b. Which construction tasks can be done together
   c. Which construction tasks must follow other operations
   d. All of the above
   e. None of the above

7. Project schedules are necessary tools to control the _____ component of a construction project.
   a. Cost
   b. Time
   c. Quality
   d. All of the above
   e. None of the above

8. Which of the below listed statement(s) is true about bar charts?
   a. They are easily readable.
   b. They are also called Gantt charts.
   c. They do not show activity interrelationships.
   d. All of the above
   e. None of the above

9. Schedules are important management tools during which stage(s) of a project?
   a. Preconstruction
   b. Construction
   c. Postconstruction
   d. All of the above
   e. None of the above

10. An activity on node network schedule can also be called a _____.
    a. Matrix schedule
    b. Linear balance diagram
    c. Precedent diagram
    d. All of the above
    e. None of the above

Exercises

1. List the activities that you would typically accomplish before going to work or to school. Estimate how long each activity will take and then determine at what time you have to get up to arrive on time. Do some activities occur concurrently? What can you do to sleep longer and still arrive on time?

2. Bring into class examples of a bar chart, network, matrix, and other scheduling examples from actual projects. How were each of these scheduling forms used? Would any other scheduling method have worked better? Did some projects use more than one method of scheduling?

CHAPTER

8

Network Construction

From studying this chapter, you will learn:

1. The steps involved in the planning stage of a project
2. The basics of both arrow and precedent notation
3. How to construct a network diagram
4. How to number and formally present a network diagram

Introduction

Network Definition

A network schedule is a logical and ordered sequence of events that describes in graphical form the approach that will be taken to complete the project.

Why Network Schedules?

A network based schedule provides the level of detail required to manage the planning, development, and construction of sophisticated projects. Although used extensively by the construction industry, network based schedules are used in many other enterprises as well. Anytime that a team of people is required to plan, organize, and manage a complicated process, a network schedule could profitably be used. In construction most owners now require contractors to utilize network schedules on midsize and major projects.

The development of a network based schedule follows a logical process that should involve most of the key project participants. A key project participant would be defined as any project team member who represents the owner, designer, contractor, or construction manager, or any consultant who has been brought aboard during the critical early stages of a project. Every network schedule must begin with a complete understanding of the project objective, which is defined by the owner. The project participants then take this objective and break it down into definable tasks that provide the level of detail necessary to organize, manage, and control the process. Once the tasks, also called **activities,** are defined, the project team organizes or "builds" the project on paper, establishing the order of the tasks to be performed in a way that allows the project to be completed in the most efficient manner. The most efficient manner may not necessarily be the fastest. The developed network may define an approach that minimizes impact on existing operations, or one that represents the best cash flow for the owner. These project goals should be continually incorporated as the team assembles the network schedule. Team members need to decide what activities must precede, succeed, or run concurrently with other activities. The end result of this process is a network, or logic diagram that accurately depicts the order that the work will follow.

A key part of this process that will pay great dividends later in the project is to get everyone to "buy into" the network schedule. If everyone involved in the project agrees and subsequently supports the logic established, the project team will experience a greater level of cooperation later on when the project is under construction. Suppliers, subcontractors, and project managers will all be in a position to understand each other's concerns and will be more willing to schedule their work to help each other.

The Project Planning Process

Project Investigation

As stated previously, the preparation of a network schedule begins with a complete understanding of the project's objective by all project team members. This is an investigatory process which requires extensive knowledge of the work environment in which the project will be constructed, the language and intent of the contract, and the scope of the project as described in the drawings and specifications.

Work Environment

The information shown on the Job Site Analysis form that was shown in Figure 6-1 is typical of what the project team should gather about the work environment before beginning to undertake a network diagram for a project. As an example, for the renovation of a hospital the scheduling team would need to determine the extent of dust protection required, allowable hours for work, points of access for workers, location and timing for equipment deliveries, and the extent of hazardous waste expected to be encountered, to name but a few concerns. The scheduler must ascertain from the owner when work can begin and when the project must be completed. In many projects there are key coordination points where certain parts or phases of the project must be complete. These points in the project are called **milestones;** they occur in response to important outside events or agreements that impact most projects.

Weather, site conditions, and unique area considerations must be determined at this stage of a project. The author was involved in a construction project that was

Figure 8–1
As the truss arrives it is immediately lifted into place to keep the work site free and clear.
Courtesy of New England Deaconess Hospital and Walsh Brothers, Inc.
Photo by Don Farrell

adjacent to an active Air Force Titan launch facility. On the days of a launch it was necessary to vacate the job site for several hours. The exact days of the launches were not known in advance, but the number of launches in a year were predictable. These had to be considered in the schedule for that particular project. The availability of building materials, access for deliveries, availability of power, and the amount of construction in the region (since other projects draw on the same labor and material suppliers) are all factors that must be understood and reflected in the schedule. For example, the job site of a project that is constructed in a dense urban environment must be tightly controlled so that the delivery of materials coincides tightly with the schedule for installation. Materials left sitting on the site are susceptible to vandalism or theft and can cause congestion in an already limited work area (see Fig. 8–1).

Contract

The complete understanding of the contract by all project participants is essential. The contract describes the responsibilities of each participant to the project. The contract will identify what materials and work will be provided by the owner. It is not unusual for the owner to prepurchase materials that may take a long time to be fabricated and delivered and provide these to the contractor, thereby saving project time. These items are called "long-lead" items; by developing a network schedule an owner is able to identify which items it makes sense to prepurchase in this way. Some contracts include incentive clauses as well as liquidated damage clauses, which provide bonuses or penalties respectively to contractors who finish the project early or late (see Santa Monica Freeway sidebar). Other features included in a contract

Sidebar

Santa Monica Freeway Reconstruction

On January 17, 1994 a critical stretch of the Santa Monica freeway was destroyed by what was termed the "Northridge Quake." This disaster removed from operation a highway which carried 290,000 cars and trucks a day. This traffic was forced onto surface streets, disrupting both interstate and local traffic terribly.

The California Department of Transportation (Caltrans), in a rush to get the highway repaired, established a very short construction period of 140 days with an incentive clause of $200,000 per each day the project was completed early. In a normal nonemergency situation the construction time for this type of project would be close to two years.

C.C. Myers Inc., a Sacramento based construction company, won the job with a base bid of $14.9 million, an amount that Myers' Executive V.P. Carl Bauer was quoted as saying was not enough to cover the inefficiencies of providing standby equipment, additional support labor, and other costs required to accelerate the construction process. Myers' gamble was to recoup its costs plus earn additional profit by finishing early and collecting the daily bonus.

By working around the clock and by efficiently scheduling operations so that equipment, materials, and people were always on hand, the Santa Monica freeway was opened 84 days after it collapsed in the quake. By bringing the job in 74 days early, the C.C. Myers Co. earned a $14.8 million bonus before adjustments for any contract changes.

---

would be specific project and milestone completion dates, work rules governing safety and hiring requirements, or any other conditions, rules, or stipulations that might affect the organization of the project.

Drawings and Specifications

The actual technical requirements of the project are covered in the drawings, specifications, and any addenda. (An addendum is a change in the technical requirements of a project that occurs after bids are let but before they are received.) These contract documents along with an estimate, if it has been prepared, must be thoroughly reviewed. The drawings and specifications are the source of all the materials and quantities used and also describe techniques and submittal requirements. If the design team is part of the scheduling process, they can clarify any unique project features, special materials, or techniques that are part of the project.

Activity Definition

Once the investigatory work is complete and all the key project team members have expressed and agreed to the major project objectives, the team is ready to begin to construct the network diagram. The first step in the networking process is to define the work breakdown structure (WBS) for the project and from that define the activities that will make up the network diagram (see WBS sidebar). An activity must be a definable part of the overall project. It must be measurable; you should be able to assign it to a project team member, and it should consume project time and resources.

Construction activities can generally be characterized as being one of three types: production, procurement, or administrative.

Production Activities

Production activities are the activities that define the actual physical construction of the project. Examples might be Erect Steel Stud Wall, Run Electrical Conduit, or Install Drywall. If the project is composed of multiple floors, or project phases, the activity would designate the activity location, such as Run Electrical Conduit, 1st Floor. The activity should be large enough to identify a meaningful quantity of work, yet small enough to allow the scheduler to sort the activity by assigned trade and by project location. A well developed schedule allows the project team to separate out all activities of each trade in such a way that each contractor has a clearly defined scope of work without overlap or omission. This is done by ensuring that each activity is the responsibility of one trade only.

The production portion of the schedule should also be sortable by floor and project phase. This ability to break out the project provides a tremendous control tool for management as the project is prepared for construction.

Procurement Activities

Activities can also be categorized as procurement or purchasing activities. These are the activities that need to occur to get all of the materials, equipment, and subcontractors to the job site. It is not unusual for all of these activities to be grouped into a separate procurement schedule, which would be managed by the purchasing department. Examples of purchasing activities would be Order Tile, Approve Roofing Sample, or Prepare Structural Steel Connection Shop Drawing.

As identified in Figure 8–2, most procurement activities follow a logical sequence beginning with the preparation of the submittal by the supplier or subcontractor, the approval of the submittal by the owner and/or designer, the ordering and fabrication of the work item, and then the delivery of the item to the job site.

In managing the procurement part of the schedule, some allowances should be made for the fact that not all submittals are approved at the first submittal. Many submittals must be resubmitted and therefore reviewed again. This can be a time-consuming process, and if the material involved is an item with a long delivery time

Sidebar

Work Breakdown Structure

Most modern day construction projects are designed, organized, and built by teams of specialized professionals. To efficiently organize this process it is necessary to break down the project into specific parts that can be coordinated and controlled. The manner in which this project is structured is termed the project's work breakdown structure.

Once the work breakdown structure for the project is established the project will be designed, budgeted, and controlled in accordance with this system. In establishing the project's work breakdown structure, each work package must be clearly defined using a written verbal description (scope of work) as well as information as to what work will be included with this part of the project. Think of the entire project as a pie—each work package defines a piece of the pie, with the goal of the project's work breakdown structure being that no overlap or omission occur (see Fig A).

Figure A
Project work breakdown.

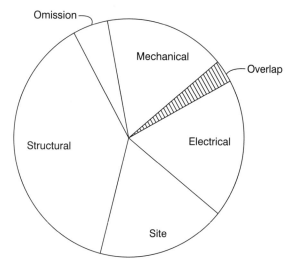

The work breakdown structure for the project needs to be established early if the project is to be fast-tracked. This allows a master schedule to be developed that enables the project manager to coordinate the design, bid, and construction phases of the project. Work that can be constructed early would be designed first and bid out. To allow this work to begin, design guidelines must be established to guarantee that constructed work is adequate to accommodate future design decisions. In most cases cost and schedule targets will be established for each work package, allowing project management to better monitor and forecast project performance.

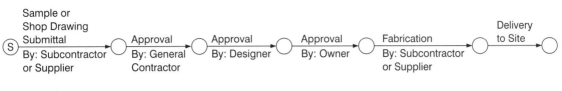

Figure 8–2
Typical procurement sequence.

this resubmittal may delay the project. On a project that the author was involved in, this process was streamlined by physically locating the structural designers in the office of the structural steel fabricator/contractor. The structural steel work on this project was critical; by improving communication between the designers and the contractors, the submittal process became faster and more efficient.

In developing the procurement schedule it is important to identify all of the required submittals for the job, as well as which offices and agencies review the submittal. As an example, an electrical supplier would submit to an electrical subcontractor, who would then submit to the general contractor. After this review the submittal would go to the owner, who would then copy the electrical designer as well as possibly the end user (tenant) and the construction manager. Any of these reviewers could comment on this submittal and possibly force a resubmittal. As the developer of the network schedule, it is important to verify the agencies that have formal review authority over the submittal and identify which reviews will be occurring concurrently and which sequentially. Both the submittals required and the time frame allowed for review should be identified in the specifications. Because the delivery of long-lead items such as structural steel, elevators, and special equipment such as large-diameter water valves or compressors can all affect the length of a project, owners often begin the submittal process before a contractor is selected.

Administrative Activities

The third category of activity that appears in a network schedule is the administrative activity. Examples of these are required inspections by local officials or by federal or regulatory agencies. This category also includes activities that occur in the permitting process, such as presentations before an architectural review commission or a zoning board of appeals. These are key steps in the life of a project that must be scheduled. Identifying these activities is going to require the involvement of all the principal parties, since normally no one party knows all of the administrative steps that have to be followed. The construction team would be able to define the construction-related inspections, such as rough and finish plumbing inspections. The designer would identify any design reviews by third parties, such as those for zoning variances or by historical commissions or by neighborhood groups. The owner would be principally involved in working with tenants and in securing financing for the project. The owner would define any major coordinating steps that would have to take place with these people.

Milestones

A key event in the life of a project can be identified by the use of a milestone date or activity. The prearranged shutdown of plant operations, the delivery of a certain phase of a building to a tenant, or a city agreement that a road be opened by a specific date are all examples of milestones. Unlike an activity, a milestone cannot be assigned to a company or person and does not consume time or resources. A milestone is used to signify an important point in the life of a project; the milestones of a project are often developed into a separate schedule (see Fig. 8–3).

Milestones are also used to "flag" significant project accomplishments serving as a measure of project success. Examples of these may be the erection of the last piece of steel, the closing in of a building (now weathertight), or the completion of a certain phase of a project.

The Network Diagram

A network diagram is a pictorial representation of the activities and the order in which they must occur to complete the project in the most efficient manner. The process of developing the diagram provides those involved with the opportunity to think through the project by constantly asking the questions:

1. What activity must occur before this activity can be done?
2. What activity must follow this activity?
3. What activity can be accomplished at the same time that this activity is occurring?

It must be understood that the order in which activities occur on a project has a certain degree of flexibility about it, and that based on personal preference and past experiences different contractors will work a project in different ways. The key, however, to the well-managed project is that when the network diagram is completed all project participants agree that the project will be run as diagramed. If differences of opinion exist, this is the place to resolve them. It is a lot easier to reorganize the job on paper than it is in the field when all suppliers and subs are under contract.

Network diagrams are constructed in accordance with a number of conventions:

1. All networks have a single starting point and finish point.
2. Networks are continuous. That is, each activity—except the first and last—has both preceding and succeeding activities.
3. In arrow notation no activity/operation can start until all preceding operations have been completed. In node notation activities can be linked by establishing relationships between the starts and finishes of the respective activities.

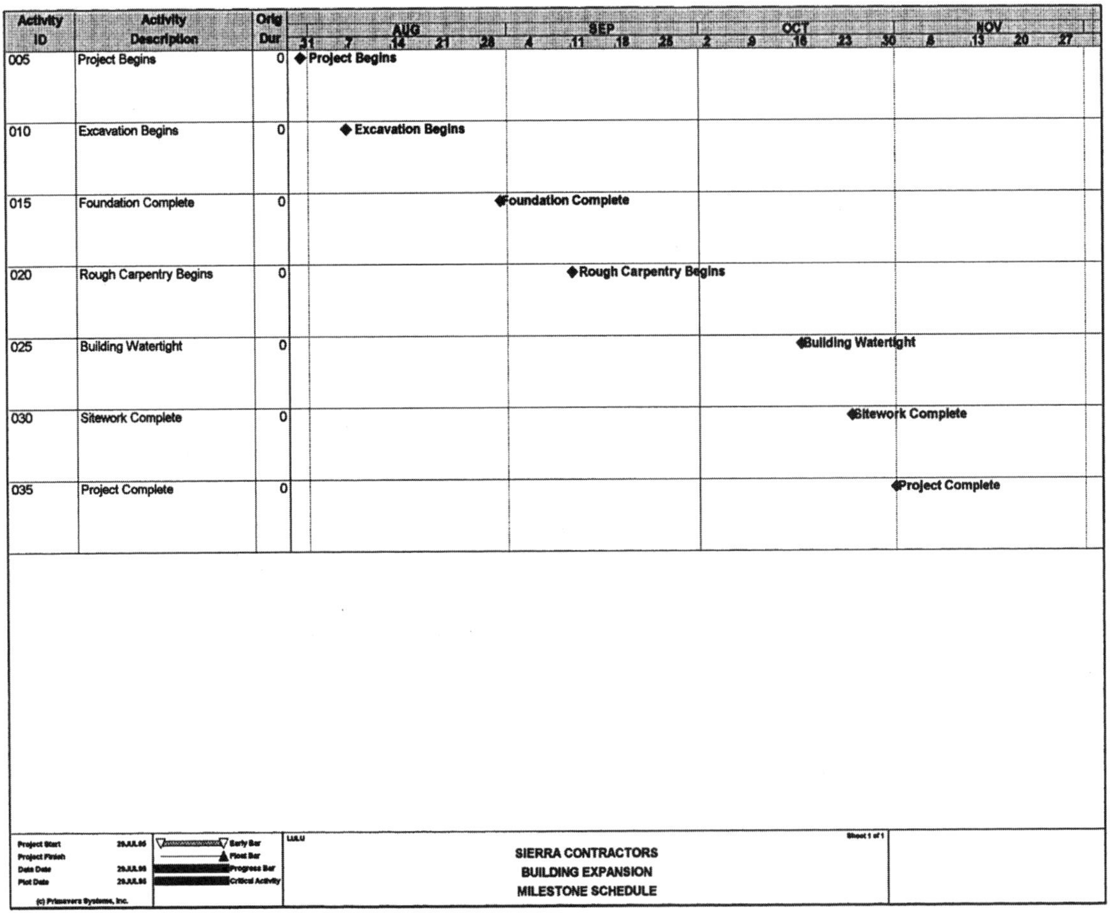

Activity ID	Activity Description	Orig Dur	AUG					SEP				OCT				NOV		
			31	7	14	21	28	4	11	18	25	2	9	16	23	30	6	13	20	27
005	Project Begins	0	◆ Project Begins
010	Excavation Begins	0	◆ Excavation Begins
015	Foundation Complete	0						◆Foundation Complete
020	Rough Carpentry Begins	0							◆Rough Carpentry Begins
025	Building Watertight	0												◆Building Watertight
030	Sitework Complete	0												◆Sitework Complete
035	Project Complete	0													◆Project Complete

Project Start	29JUL96	▽▬▬▬▬▽ Early Bar	U/U		Sheet 1 of 1
Project Finish		▬▬▬▬▲ Float Bar		SIERRA CONTRACTORS	
Data Date	29JUL96	▬▬▬ Progress Bar		BUILDING EXPANSION	
Plot Date	29JUL96	▬▬▬ Critical Activity		MILESTONE SCHEDULE	
(c) Primavera Systems, Inc.					

Figure 8–3
Milestone schedule.
This graphic was created using Primavera Project Planner® (P3®), a product of Primavera Systems, Inc.

4. The arrows in an arrow diagram are not drawn to scale; meaning that the length that the arrow is drawn on the diagram does not indicate its duration.

5. Each arrow or node indicates a single activity.

A point of debate both in industry and academia is whether the network diagram should be developed in arrow notation or precedent notation (see Fig. 8–4). As can be seen, in arrow notation the activity, or work, takes place on the arrow, while in precedent notation it occurs on the node. Technically, the results of both notations provide the same results, although more and more professionals are beginning to work in precedent notation because of its greater flexibility.

Activity on Arrow

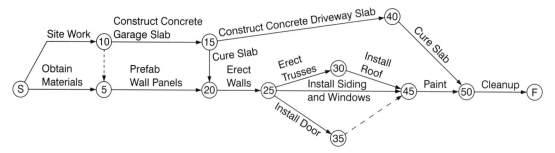

Activity on Node

(Note: Finish to start lag of 24 days used instead of curing activity above.)

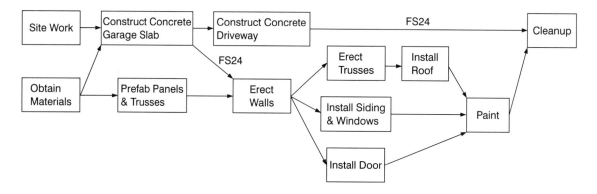

Primavera for Windows

(Time-scaled bar — activities are visually connected on screen.)

Activity ID	Activity Description	Orig Dur	SEP	OCT	NOV	DEC	JAN	F
1	SITE WORK	10	▄SITE WORK					
2	OBTAIN MATERIALS	8	▄▼OBTAIN MATERIALS					
4	PREFAB WALL PANLS.&	16	▄▽─────────────▼PREFAB WALL PANLS.& ROOF TRUSSES					
3	CONSTRUCT CONC.	6	▄▼CONSTRUCT CONC. GARAGE SLAB					
12	CONCRETE DRIVEWAY	8	▄▽──────────▼CONCRETE DRIVEWAY SLAB					
6	ERECT WALLS	4		▄▼ERECT WALLS				
7	ERECT TRUSSES	4		▄▼ERECT TRUSSES				
8	INSTALL SIDING &	10		▄▽▼INSTALL SIDING & WINDOWS				
9	INSTALL DOOR	4		▄▼INSTALL DOOR				
10	INSTALL ROOFING	12		▄▼INSTALL ROOFING				
11	PAINT	16		▄▼PAINT				
14	CLEANUP	4		▄▼CLEANUP				

Figure 8–4
Example network solutions.

No matter what diagraming method is used, the key to successful implementation is for all participants to understand it, participate in its development, and fully utilize it once construction begins. In this book both arrow and precedent diagraming methods will be covered and utilized in examples.

Activity on Arrow Notation

As can be seen in the art, in arrow notation the activities are identified by arrows which are connected by nodes. Each activity is represented by a single arrow, which is read from left to right. In a pure logic diagram the arrow is not drawn to scale. The purpose of the diagram is to establish the logic of the project, identifying the order of the tasks. The diagram starts at a single starting point and ends at a single finish point. The junctions between arrows are called **nodes;** by definition they consume neither time nor resources. A few types of logic statements and their diagrams are shown in Figure 8–5.

Activity on arrow notation is also called *i-j* notation; this name was established at about the time that CPM was developed. In that notation the *i* node would be the first node at the tail of the arrow and the *j* node would be the second node at the point of the arrow. These numbers are important since they form addresses for each of the arrows in the network. For example, in Figure 8–6 the Rough Plumbing activity would be described as (25–35) using *i-j* notation.

Figure 8–5
Logic statements: activity on arrow notation.

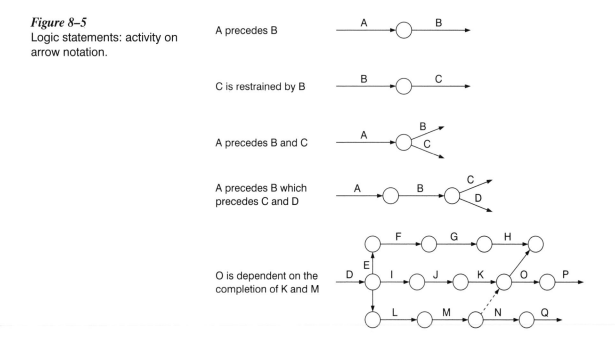

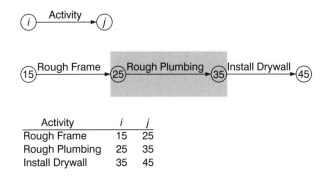

Figure 8–6
i-j notation.

Activity	*i*	*j*
Rough Frame	15	25
Rough Plumbing	25	35
Install Drywall	35	45

The arrows that define activities are solid arrows and should be titled or labeled in such a way that anyone involved with the project can look at the activity and see how the activity relates to other activities in the project. The use of abbreviations is acceptable provided others involved in the project understand the abbreviation used. It is not a good idea to use numbers or other codes since that will require the reader to continually refer to a legend and will hinder future discussion about the logic shown.

Activity on arrow diagrams may require the use of a dashed arrow called a **dummy.** A dummy arrow establishes a relationship between two activities without adding another activity. A dummy has no duration and consumes no resources. A dummy arrow is necessary for the following reasons:

1.   In many instances the logic of a project cannot be shown correctly without the use of dummy arrows.

Consider the network shown in Figure 8–7.

Let us assume that each of these activities is accomplished by a different crew and that we now must add a second wall to the project which will be handled by the same crews. Which diagram is correct, Figure 8–8a or 8–8b?

The reader sees in Figure 8–8a that the installation of the steel stud wall for wall 2 must be completed before the drywall for wall 1 can occur. Clearly this is not true, so what do we do? This demonstrates one of the uses of a dummy arrow. A dummy arrow can be viewed strictly as a dependency; it adds no additional work, it only defines an additional relationship. Figure 8–8b shows the project drawn correctly with the use of a dummy arrow.

2.   Dummy arrows also allow the scheduler to provide unique *i-j* addresses for project activities (see Fig. 8–9).

One can see from Figure 8–9a that the Site Layout activity and the Mobilize Equipment activity have the same *i-j* numbers. This problem can be solved by using

Figure 8–7
Part of a sample network.

Install Steel Stud Wall → Insulate Wall → Install Drywall →

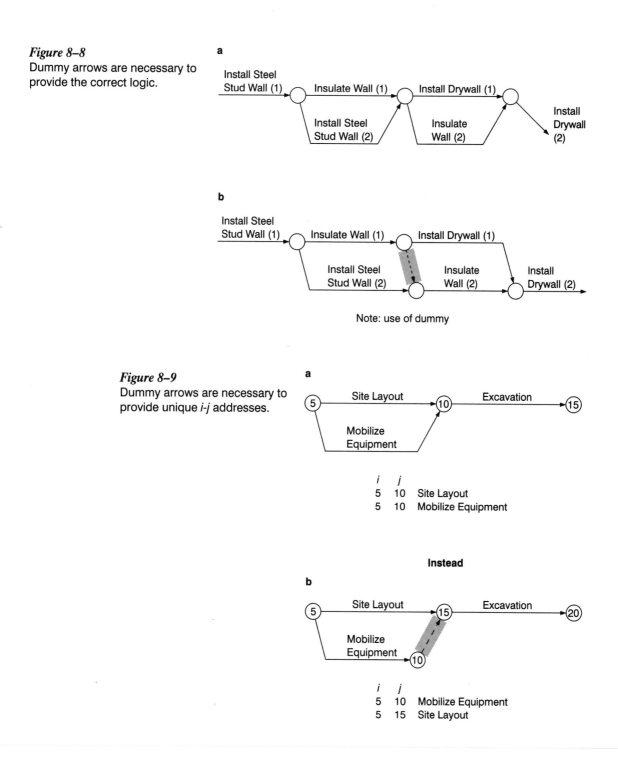

Figure 8–8
Dummy arrows are necessary to provide the correct logic.

a

Install Steel Stud Wall (1)
Insulate Wall (1)
Install Drywall (1)

Install Steel Stud Wall (2)
Insulate Wall (2)
Install Drywall (2)

b

Install Steel Stud Wall (1)
Insulate Wall (1)
Install Drywall (1)

Install Steel Stud Wall (2)
Insulate Wall (2)
Install Drywall (2)

Note: use of dummy

Figure 8–9
Dummy arrows are necessary to provide unique *i-j* addresses.

a

⑤ — Site Layout — ⑩ — Excavation — ⑮

Mobilize Equipment

i	*j*	
5	10	Site Layout
5	10	Mobilize Equipment

Instead

b

⑤ — Site Layout — ⑮ — Excavation — ⑳

Mobilize Equipment — ⑩

i	*j*	
5	10	Mobilize Equipment
5	15	Site Layout

205

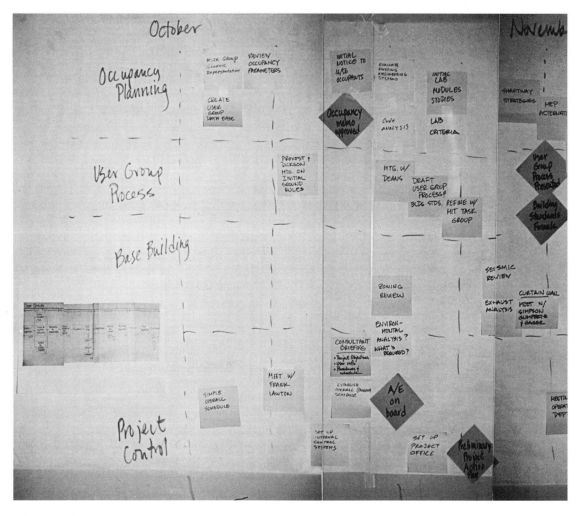

Figure 8–10
"Post-it notes" are often used to build a network's logic on a wall.
Photo courtesy of Beacon Construction

a dummy arrow which creates an additional node, allowing each activity to have its own unique address as shown in 8–9b.

3.   Dummies are also used for convenience sake as the network is being drawn. Many times additional relationships need to be identified between activities in a nearly complete network diagram. Rather than redraw the entire network, a dummy arrow will be used instead to establish the necessary relationship. This quickly establishes the relationship without the addition of another activity and without taking the time necessary to redraw the network.

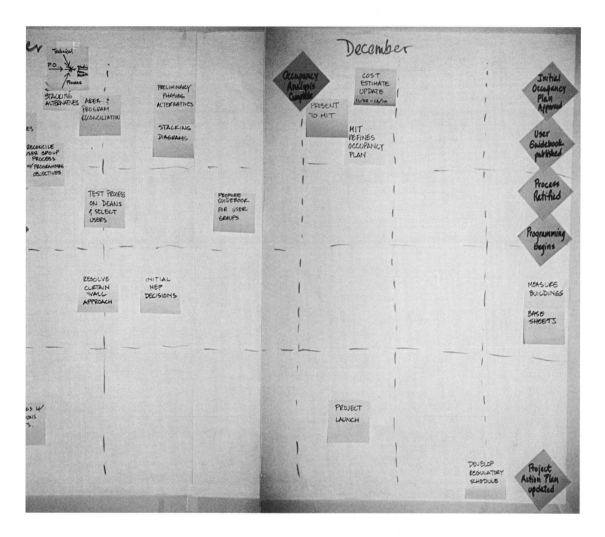

Precedent Notation

The use of precedent notation (also called activity on node notation) is common in industry. In this notation the individual operations, or activities, are placed on the nodes, which are then connected by arrows. This notation is clearly the inverse of arrow notation. A clear advantage of this notation is that dummy arrows are not necessary. The activities are placed within circles or boxes and are connected by arrows as appropriate (see Fig. 8–10).

Arrow	Precedence

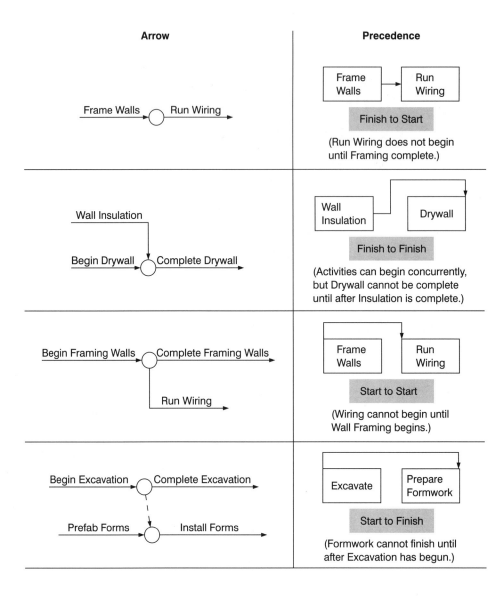

All of the above links can include a
link duration as illustrated below:

(Formwork cannot finish until
4 days after Excavation begins.)

Figure 8–11
Arrow notation compared to precedent notation.

208

A characteristic of precedent notation is the ability to overlap activities, allowing the scheduler to more accurately model the project's operations. The use of overlapping activity notation, also called **"leads and lags"**, benefits from the use of a computer to calculate the project duration and individual activity start and finish times, since the computations become very complicated. The use of "leads and lags," which is illustrated in Figure 8–11, is a recognition of the fact that project operations are not perfectly sequential and that often activities run in parallel, or are phased with certain activities being given a head start. Activity on arrow notation can handle these complexities, but requires the creation of additional activities; for example, the activity Framing may be broken down into three activities, Begin Framing, Continue Framing, and Complete Framing. This breakdown allows the integration of other activities that parallel the framing activity as can be seen in the illustration.

The same operational sequence could be shown in precedent notation by the use of Start to Start, Finish to Finish, and Start to Finish relationships. By using these kinds of connectors the scheduler is able to choose the type of relationship that most accurately represents what will really happen in the field. In arrow notation the only type of connection allowed is a Finish to Start, which indicates that the preceding activity is completely finished prior to the start of the next activity.

As illustrated at bottom in Figure 8–11, the starting of an activity can also be delayed by the use of what is called a link duration or lag. The use of a lag can be helpful in clearly modeling the anticipated job site operations. A good example might be in sequencing the placement of concrete followed by the setting of base plates. The specifications call for a 2-day curing of the concrete before the base plates can be set. In arrow notation the operations would be as shown in Figure 8–12a. In precedent notation the operations would be shown as in Figure 8–12b.

As can be seen in the art, either notation can illustrate the operations as they will occur at the job site, but precedent notation requires fewer activities and can model the job sequence faster. For this reason most companies and most software

Figure 8–12
Lag example.

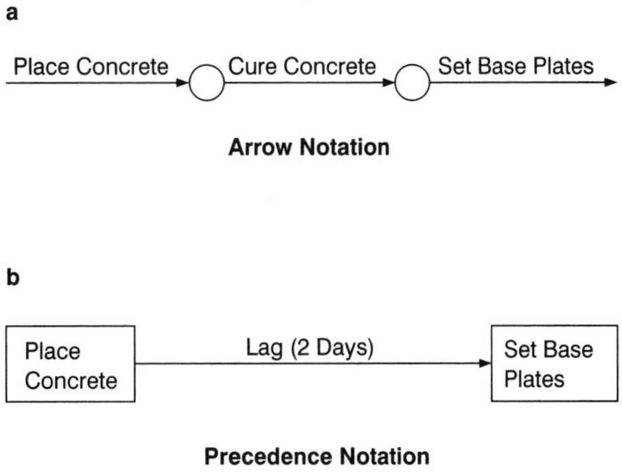

Arrow Notation

Precedence Notation

packages are now set up utilizing precedent notation. Arrow notation has the advantage of being easier to calculate by hand and more graphical; it therefore serves as a good technique to learn scheduling basics.

Network Presentation

The final step in the networking process (a step which marks the completion of the planning stage of the project) is the preparation of the final network diagram. This diagram should clearly indicate the scope of the project, as defined by the activities present in the schedule, as well as the planned sequencing of the project, as identified by the order in which the activities are connected. Again, no matter which notation is used, the sequencing of the job should be clearly illustrated by the network diagram.

It is important at this stage that all of the key project participants agree on and support the sequence in which the activities are structured. The final network diagram needs to consider major project decisions such as phased construction, planned construction techniques, long-lead purchasing items, and owner concerns such as tenant occupancy and financing requirements. Up to this point individual activity durations have yet to be considered. It is important in the planning stages that the sequencing of the individual operations not be altered because of the planned durations of other concurrent operations. It is always best to consider sequencing and activity durations independently, then "run the network," then make all the final adjustments.

Some of the final steps that are part of the presentation of the final network are to number the nodes/activities, center the key activities on the page, and generally "neaten up" the final network. If a network is being produced at this stage by hand, nodes are generally numbered from left to right and vertically from top to bottom (see Fig. 8–13). The exception is that if an activity points upward in arrow notation, its *i* node is always less than its *j* node, as shown in that illustration. It is always best to minimize crossovers, although usually they cannot be totally avoided. It is also a plus if operations of a similar nature are clustered together, for instance, procurement activities or structural steel work or groupings by building floor. Most network schedules, however, are now produced using computer software, which makes this step easier since the software will automatically sort and position the presentation as specified by the scheduler. Normally, the schedule output will be set up to organize the activities by project phase, subcontractor or discipline, responsibility, and the like, further sorted by start date or float (see Fig. 8–14).

The issues of how you want to manage the job and how you would like the information presented should be decided prior to the entering of the activity data. As activities are entered into the computer, information about the activity such as its location on the project, the trade or responsibility to do it, resource requirements, activity cost, and so on can be entered at the same time, giving the user a tremendous amount of

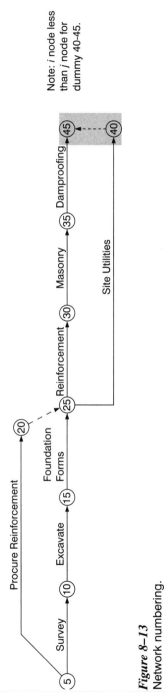

Figure 8–13
Network numbering.

Note: *i* node less than *j* node for dummy 40-45.

Act ID	Activity Description	Orig Dur	Rem Dur	%	Early Start	Early Finish	Total Float	Resource	Budgeted Cost	UL	AUG	SEP

Engineering Department

Andy Mason - Director of Development

Act ID	Activity Description	Orig Dur	Rem Dur	%	Early Start	Early Finish	Total Float	Resource	Budgeted Cost	
BA400	Design Building Addition	20	20	0	19JUL93	13AUG93	1	DES ENG	9,600.00	△▬▬▽Design Buil
BA469	Assemble Technical Data for	3	3	0	03SEP93	08SEP93	61	DES ENG	540.00	△▽
BA470	Review Technical Data on	10	10	0	09SEP93	22SEP93	61	DES ENG	600.00	△▬▽

Acme Motors - Owner

BA501	Review and Approve Designs	14	14	0	16AUG93	02SEP93	1	DES ENG	1,680.00	△▬▬▽Reviev
BA530	Review and Approve Brick	10	10	0	20SEP93	01OCT93	35	DES ENG	240.00	△▬
BA560	Review and Approve Flooring	10	10	0	20SEP93	01OCT93	120	DES ENG	240.00	△▬

Purchasing Department

Meg Foley - Purchasing Manager

BA450	Assemble Brick Samples	10	10	0	03SEP93	17SEP93	35	UNASSIGN	0.00	△▬▽
BA480	Assemble and Submit Flooring	10	10	0	03SEP93	17SEP93	120	UNASSIGN	0.00	△▬▽
BA411	Prepare and Solicit Bids for	3	3	0	23SEP93	27SEP93	61	ACCTS	384.00	△▽
BA412	Review Bids for Heat Pump	2	2	0	28SEP93	29SEP93	61	ACCTS	128.00	▨
BA413	Award Contract for Heat	1	1	0	30SEP93	30SEP93	61	ACCTS	32.00	▨
BA550	Fabricate and Deliver Heat	90	90	0	01OCT93	09FEB94	61	VENDOR	0.00	△
BA421	Prepare and Solicit Bids for	3	3	0	04OCT93	06OCT93	35	ACCTS	384.00	
BA407	Prepare and Solicit Bids for	5	5	0	04OCT93	08OCT93	120	ACCTS	640.00	
BA422	Review Bids for Brick	3	3	0	07OCT93	11OCT93	35	ACCTS	96.00	
BA408	Review Bids for Flooring	3	3	0	11OCT93	13OCT93	120	ACCTS	96.00	
BA423	Award Contract for Brick	1	1	0	12OCT93	12OCT93	35	ACCTS	32.00	
BA600	Deliver Brick	60	60	0	13OCT93	10JAN94	35	VENDOR	0.00	
BA409	Award Contract for Flooring	1	1	0	14OCT93	14OCT93	120	ACCTS	32.00	
BA620	Fabricate and Deliver Flooring	60	60	0	15OCT93	12JAN94	120	VENDOR	0.00	

Construction Department

Joe Nolan - Construction Manager

BA630	Begin Building Construction	0	0	0	03SEP93		1	UNASSIGN	0.00	◆Begin
BA640	Site Preparation	20	20	0	07SEP93*	04OCT93	0	EXCAVATR	20,640.00	△▬
BA650	Excavation	10	10	0	05OCT93	18OCT93	0	EXCAVATR	13,760.00	
BA660	Install Underground Water	5	5	0	19OCT93	25OCT93	0	PLUMBER	6,600.00	
BA670	Install Underground Electric	5	5	0	19OCT93	25OCT93	0	ELECTRCN	6,400.00	
BA680	Form/Pour Concrete Footings	10	10	0	26OCT93	08NOV93	0	RGHCARP-	63,240.00	
BA681	Concrete Foundation Walls	10	10	0	09NOV93	22NOV93	0	FNISHR+	70,920.00	
BA690	Form and Pour Slab	5	5	0	23NOV93	01DEC93	0	RGHCARP-	19,012.00	
BA700	Backfill and Compact Walls	2	2	0	02DEC93	03DEC93	0	EXCAVATR	2,144.00	
BA701	Foundation Phase Complete	0	0	0		03DEC93	0	UNASSIGN	0.00	
BA702	Begin Structural Phase	0	0	0	06DEC93		0	UNASSIGN	0.00	
BA710	Erect Structural Frame	20	20	0	06DEC93	04JAN94	0	IRWK+	111,600.00	
BA712	Floor Decking	14	14	0	05JAN94	24JAN94	0	IRWK+	70,840.00	
BA730	Concrete First and Second	15	15	0	25JAN94	14FEB94	0	LABORER+	75,540.00	
BA809	Rough-In Phase Begins	0	0	0	03FEB94		32	UNASSIGN	0.00	
BA810	Set Mechanical and Electrical	15	15	0	03FEB94	24FEB94	32	OPENG+	25,920.00	
BA720	Erect Stairwell and Elevator	10	10	0	15FEB94	01MAR94	0	LABORER	9,600.00	
BA731	Concrete Basement Slab	10	10	0	15FEB94	01MAR94	0	FNISHR+	50,360.00	

			BASE BLDG		Sheet 1A of 2B
Project Start	19JUL93	△▬▬▽ Early Bar			
Project Finish	22JUL94	▼ Float Bar	Acme Motors		
Data Date	19JUL93	△▬▬▽ Progress Bar	Plant Expansion and Modernization		
Plot Date	16JAN90	▬▬ Critical Activity	Float Bars by Department, Resp		
(c) Primavera Systems, Inc.					

Figure 8–14
Example activity breakout.
This graphic was created using Primavera Project Planner® (P3®), a product of Primavera Systems, Inc.

project information. Later in Part Four of this textbook, control applications and more specifics about work breakdown structure and activity coding will be discussed.

Conclusion

The completion of the network diagram marks the completion of the planning stage of a project and is important for a number of reasons. One, a network diagram should define the most expeditious schedule, which (as illustrated in the Santa Monica Freeway sidebar) can lead to significant cost savings on the project. Two, the schedule defines the impact of one activity on another, which can become critically important when delays occur on a project, and a work-around solution must be determined. Three, once durations are specified, the network diagram can determine the starts and finishes of each activity. This is the subject of the next chapter.

A network is a working tool. To keep it effective it must be updated and adjusted as the project moves along. The logic that has been established at the planning stage is based on the best knowledge available, but certainly as real events happen on the job site, refinements must occur. It is best to think of network scheduling as an iterative process—a scheduling team makes its best estimates of the order and duration of activities, then the schedule is run and examined. Adjustments are made and the schedule is run again. This process is repeated over and over again even as construction begins. The only difference is that real durations and events begin to play a role in the iterations.

Chapter Review Questions

1. In arrow notation the longer the arrow, the longer the duration of the activity.

    ___ T   ___ F

2. Each activity should have its own unique *i-j* address.

    ___ T   ___ F

3. The creation of a network schedule is an important component of the planning stage of a project.

    ___ T   ___ F

4. It is acceptable to use a single arrow to represent multiple activities.

    ___ T   ___ F

5. The convention used to number networks from left to right and vertically is done to provide a technically correct solution.

    ___ T   ___ F

6. Given the diagram below, which of the following statements is correct?

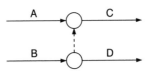

   a. Activity C can begin once activity A is completed.
   b. Activity D can begin once activities A and B are completed.
   c. Activity C can begin once activities A and B are completed.
   d. None of the above are true.

7. Which of the following is not a typical activity category?
   a. Production
   b. Setup
   c. Procurement
   d. Administrative

8. In activity on arrow notation the junction between arrows is called a:
   a. Node
   b. Event
   c. Milestone
   d. A and B are both correct
   e. None of the above

9. A dummy arrow is necessary for which of the following reasons?
   a. Necessary to show correct logic
   b. Allows unique i-j addresses
   c. Convenient in drawing and adjusting network diagrams
   d. All of the above

10. Certain key events are called:
   a. Dummies
   b. Critical events
   c. Nodes
   d. Milestones

Exercises

1. Draw the logic diagram (use both activity on arrow and precedent notation) for the following set of activities:

A, B, D, and K start the project
C follows A and B
E and F follow C and D
E precedes G

2. Draw the logic diagram (use both activity on arrow and precedent notation)
   for the following set of activities:

   A and B precede D
   C is restrained by A
   E and D precede F and G
   F follows C
   K follows F, G, and H
   G and H precede I

Activity Duration and Network Calculations

CHAPTER OUTLINE

From studying this chapter, you will learn:

1. How to determine the duration of an activity

2. The definitions and importance of *early start, early finish, late start, late finish,* and *total float* times

3. How to calculate each of those times

4. The definition of a *critical path* and how to calculate it

Introduction

One of the first questions an owner or project manager wants answered is "When can the project be completed?" Subcontractors and vendors need to know when their work can start and when to schedule delivery of their materials. If work needs to be delayed because of outside constraints, management needs to know how long the work can be delayed without impacting the project.

To determine the completion date of a project it is first necessary to ascertain the relationships and durations of each of the individual activities. When a completion date for a project is calculated, it is assumed that the activities will be conducted in the order identified on the network diagram and within the duration noted. It is also assumed that the project will be completed in the most efficient manner and in the least amount of time possible without working second shifts, overtime, or with large inefficient crews. To calculate the project's duration it is also necessary to consider the calendar that the job will follow. This calendar reflects weather, holidays, vacation time, and periods when other work rules apply. Once the calendar for the project is determined and the individual activity durations are set, the overall project and activity start and completion dates can be calculated. **Float** time, which reflects extra time, can also be determined at this time. These dates become critical to the project management team. All involved now know when each project activity must start and be finished and which activities can slip without affecting the overall project completion date.

In many cases the start dates and durations that are first calculated will not be acceptable to some of the project participants. Perhaps the overall duration is too long, or certain resources are overcommitted, or the start dates are not acceptable. Whatever the reason, this is the time when adjustments are made. First the team will look at available float. This will be used to adjust the project where possible. If further changes are necessary, then activities may have to be conducted in less efficient ways to accelerate the project. This may mean working overtime or second shifts, or using larger than normal crews. In summary, the process is to first calculate individual activity durations, conduct the network calculation, and analyze, and then adjust and recalculate as necessary. The process begins with activity durations, so that will be explained first.

Durations

Activity durations can be ascertained at the time that the scheduling team is involved with the development of the network diagram. If an estimate for the project is in progress, it would make sense to determine activity durations at the same time. To put together an estimate it is essential to have knowledge of individual activity times. The problem that often occurs is that the estimate line items are worked using different activities than the ones used in scheduling. To adjust for this, estimate items may have to be combined to create schedule activities. See the example below:

Estimate Line Items	*Schedule Activity*
1. Hollow Metal Frames	1. Install H.M. Frames,
2. Hollow Metal Doors	Doors, and Hardware
3. Door Hinges	
4. Locksets	
5. Panic Devices	
6. Thresholds	

As companies become more automated and the estimating and scheduling departments communicate better, the transition between the estimate and schedule can be smoothed and increased project efficiency can be achieved. This topic will be explained further in Section Four.

Depending on the type of activity, the method used to determine the duration of an activity varies. Production activities such as steel erection, rough wiring, or wood framing can be determined by talking to the appropriate subcontractor or in-house supervisor, researching past projects of a similar nature, or by looking at national data books. The duration of vendor items such as the manufacture and delivery of curtainwalls, special order cabinetry, or custom windows is best determined by talking to the particular vendor. The project's contract documents also have to be reviewed to allow for the time involved in the preparation of submittals, submittal reviews, and resubmittals if necessary. The durations of administrative activities such as securing permits or other required preconstruction approvals can be arrived at by talking to the particular agency that conducts the service. Since many of these permits are granted by government agencies, the submittal requirements, notification procedures, and time allowed for governmental action are all published information. Owners, designers, or constructions professionals who have worked on past projects in the area can also be consulted, as they have a good sense of the times and procedures involved in these administrative steps.

In figuring the duration of an activity it is important to picture the way the activity will be conducted at the job site. As an example, running electrical wire while standing on the floor can be done much faster than when working on a ladder. Work that is done as part of new construction can usually occur much more efficiently than in a renovation project while working around existing operations. Once the job conditions are understood, consider each operation as using a normal crew size, working a normal shift, and working under normal weather conditions. It is best to first work all the activities this way. After the project's activity start and finish times are calculated, adjustments can be made if necessary. By adjusting to less efficient operations afterwards, only when necessary, the project will end up being run as efficiently as possible.

Weather and other job site factors can be treated in one of two ways depending on the size of the project. For large projects that will occur over several years, it is best to create a master calendar that accounts for the normal number of good working days per month based on historical or contract conditions. As was mentioned earlier, the author was involved in a project where it was known that the job site would be shut down for Titan launches a known number of times per year, but the specific dates were classified. To account for this, random days equal to the number of launches were eliminated as work days on the master calendar. Weather is generally more seasonal, so bad weather days are generally eliminated from the calendar on a monthly basis (see Fig. 9–1).

On union projects it is not uncommon for each of the trades to have its own work rules. For instance the electrical union may have one half-day per week, say Thursday, for check cashing. Another trade may have one Friday a month off as a personal day. To account for this, computer software packages such as Primavera Project Planner allow the scheduler to create several calendars. As activities are entered into the computer they are assigned to a particular calendar.

Multiple calendars are also useful in adjusting the differing times between production activities and procurement. Production activities operate on a 5-day work week, whereas procurement or administrative activities are calculated on a 7-day week. For instance, when a window vendor states that the windows will arrive in 6 weeks, that might mean 42 days, but when a window installer says it will take 6 weeks to install all the windows, it might mean 30 days. Some production activities have similar disparities. Concrete cures on a 7-day calendar, but the concrete finishing would be worked on a 5-day calendar.

The other way to determine adjustments for weather and other job site factors is to lengthen each activity's duration accordingly. This makes most sense on a small job of short duration. If a site work activity with perfect weather conditions would take 5 days, the duration might be entered as 6 to allow for one day of bad weather. This approach works when the scheduler knows in advance during which month each activity will occur. On multiple-year projects, however, it is not always apparent in which month some of the middle activities will occur until the first calculations occur.

If experienced subcontractors or field personnel are not available to consult, then historical data can also be used to calculate activity durations. Figure 9–2, taken from Means' Building Cost Data book, shows a typical record that could be used to determine the duration for an activity. As was explained in the estimating section of this book, the Means data is based on national averages and approaches to the construc-

APRIL						1995
SUN	MON	TUE	WED	THUR	FRI	SAT
						1
2	3	4	5	6	7	8
	WP-1	WP-2	WP-3	WP-4	WP-5	
9	10	11	12	13	14	15
	WP-6	WP-7	WP-8	WP-9	WP-10	
16	17	18	19	20	21	22
	WP-11	WP-12	WP-13	WP-14	WP-15	
23	24	25	26	27	28	29
	WP-16	WP-17	WP-18	WP-19	WP-20	
30						

MAY						1995
SUN	MON	TUE	WED	THUR	FRI	SAT
	1	2	3	4	5	6
	WP-21	WP-22	WP-23	WP-24	WP-25	
7	8	9	10	11	12	13
	WP-26	WP-27	WP-28	WP-29	WP-30	
14	15	16	17	18	19	20
	WP-31	WP-32	WP-33	WP-34	WP-35	
21	22	23	24	25	26	27
	WP-36	WP-37	WP-38	WP-39	WP-40	
28	29	30	31			
	WP-41	WP-42	WP-43			

JUNE						1995
SUN	MON	TUE	WED	THUR	FRI	SAT
				1	2	3
				WP-44	WP-45	
4	5	6	7	8	9	10
	WP-46	WP-47	WP-48	WP-49	WP-50	
11	12	13	14	15	16	17
	WP-51	WP-52	WP-53	WP-54	WP-55	
18	19	20	21	22	23	24
	WP-56	WP-57	WP-58	WP-59	WP-60	
25	26	27	28	29	30	
	WP-61	WP-62	WP-63	WP-64	WP-65	

Figure 9–1
Master calendar, illustrating setup of job and work week.
This graphic was created using Primavera Project Planner® (P3®), a product of Primavera Systems, Inc.

tion project. Just as when using the Means data for estimating, the appropriate line number must be found and the correct units of measure must be considered.

To utilize the information from a historic data resource, the scheduler must know the quantity of work required. If an estimate has been done for the project, this quantity information could be taken from it. However, remember that the estimate may be using a different activity breakdown, so line items may have to be combined or separated. If an estimate has not been prepared, a separate quantity takeoff will need to be done. In the Means data (Figure 9–2) the crew, crew output, and man-hours columns supply the information that is needed to calculate an activity's duration. The crew listed for the line number is the size of crew that would most likely be assigned to accomplish that task. The standard Means crews are identified in the back of the book. The crew output column identifies how many units of work that crew can accomplish in a typical workday. Therefore if you were using that same crew:

Quantity/Crew output = Duration (in days)

032 | Concrete Reinforcement

032 100 | Reinforcing Steel

			CREW	DAILY OUTPUT	MAN-HOURS	UNIT	1994 BARE COSTS				TOTAL INCL O&P	
							MAT.	LABOR	EQUIP.	TOTAL		
109	1800	#18 bars	C-5	100	.560	Ea.	26.50	14.40	4.92	45.82	59.50	109
	2100	#11 to #18 & #14 to #18 transition		100	.560		28.50	14.40	4.92	47.82	61	
	2400	Bent bars, #10 & #11		140	.400		25	10.30	3.51	38.81	48.50	
	2500	#14		120	.467		33	12	4.10	49.10	61	
	2600	#18		90	.622		48	16	5.45	69.45	86	
	2800	#11 to #14 transition		100	.560		34	14.40	4.92	53.32	67.50	
	2900	#11 to #18 & #14 to #18 transition		90	.622		48	16	5.45	69.45	86	

032 200 | Welded Wire Fabric

			CREW	DAILY OUTPUT	MAN-HOURS	UNIT	MAT.	LABOR	EQUIP.	TOTAL	TOTAL INCL O&P	
207	0010	WELDED WIRE FABRIC Rolls, 6 x 6 - W1.4 x W1.4 (10 x 10) 21 lb.	2 Rodm	35	.457	C.S.F.	7	12.05		19.05	29	207
	0100	Sheets										
	0200	6 x 6 - #8/8 (W2.1/W2.1) 30 lb. per C.S.F.	2 Rodm	31	.516	C.S.F.	9	13.65		22.65	34	
	0300	6 x 6 - W2.9 x W2.9 (6 x 6) 42 lb. per C.S.F.		29	.552		12.90	14.55		27.45	40	
	0400	6 x 6 - W4 x W4 (4 x 4) 58 lb. per C.S.F.		27	.593		17.25	15.65		32.90	46.50	
	0500	4 x 4 - W1.4 x W1.4 (10 x 10) 31 lb. per C.S.F.		31	.516		10.30	13.65		23.95	35.50	
	0600	4 x 4 - W2.1 x W2.1 (8 x 8) 44 lb. per C.S.F.		29	.552		12.50	14.55		27.05	40	
	0650	4 x 4 - W2.9 x W2.9 (6 x 6) 61 lb. per C.S.F.		27	.593		18.15	15.65		33.80	47.50	
	0700	4 x 4 - W4 x W4 (4 x 4) 85 lb. per C.S.F.		25	.640		26.50	16.90		43.40	59	
	0750	Rolls										
	0800	2 x 2 - #14 galv. @ 21 lb., beam & column wrap	2 Rodm	6.50	2.462	C.S.F.	12.75	65		77.75	129	
	0900	2 x 2 - #12 galv. for gunite reinforcing	"	6.50	2.462	"	16.75	65		81.75	133	
	0950	Material prices for above include 10% lap										
	1000	Specially fabricated heavier gauges in sheets	4 Rodm	50	.640	C.S.F.		16.90		16.90	30	
	1010	Material only, minimum				Ton	550			550	605	
	1020	Average					700			700	770	
	1030	Maximum					855			855	940	
240	0010	FIBROUS REINFORCING										240
	0100	Synthetic fibers				Lb.	2.75			2.75	3.03	
	0110	Synthetic fibers, 1-1/2 lb. per C.Y., add to concrete				C.Y.	4.25			4.25	4.68	
	0150	Steel fibers				Lb.	.45			.45	.50	
	0160	Steel fibers, 50 lb. per C.Y., add to concrete				C.Y.	23			23	25.50	
	0170	Steel fibers, 75 lb. per C.Y., add to concrete					34.50			34.50	38	
	0180	Steel fibers, 100 lb. per C.Y., add to concrete					45			45	49.50	

032 300 | Stressing Tendons

			CREW	DAILY OUTPUT	MAN-HOURS	UNIT	MAT.	LABOR	EQUIP.	TOTAL	TOTAL INCL O&P	
307	0010	PRESTRESSING STEEL Post-tensioned in field										307
	0020											
	0100	Grouted strand, 50' span, 100 kip	C-3	1,200	.053	Lb.	1.72	1.30	.13	3.15	4.26	
	0150	300 kip		2,700	.024		1.39	.58	.06	2.03	2.59	
	0300	100' span, grouted, 100 kip		1,700	.038		1.14	.92	.09	2.15	2.92	
	0350	300 kip		3,200	.020		1.01	.49	.05	1.55	2	
	0500	200' span, grouted, 100 kip		2,700	.024		1.06	.58	.06	1.70	2.23	
	0550	300 kip		3,500	.018		.98	.45	.05	1.48	1.89	
	0800	Grouted bars, 50' span, 42 kip		2,600	.025		1.14	.60	.06	1.80	2.35	
	0850	143 kip		3,200	.020		1	.49	.05	1.54	1.99	
	1000	75' span, grouted, 42 kip		3,200	.020		.93	.49	.05	1.47	1.91	
	1050	143 kip		4,200	.015		.88	.37	.04	1.29	1.64	
	1200	Ungrouted strand, 50' span, 100 kip	C-4	1,275	.025		1.11	.68	.03	1.82	2.45	
	1250	200 kip		1,475	.022		1.12	.58	.02	1.72	2.29	
	1400	100' span, ungrouted, 100 kip		1,500	.021		.94	.57	.02	1.53	2.08	
	1450	200 kip		1,650	.019		.96	.52	.02	1.50	2	
	1600	200' span, ungrouted, 100 kip		1,500	.021		.88	.57	.02	1.47	2.02	
	1650	200 kip		1,700	.019		.89	.51	.02	1.42	1.90	

Figure 9–2

Typical source for computing activity duration.

From *Means Building Construction Cost Data 1995*. Copyright R. S. Means Co., Inc., Kingston, MA, 617-585-7880, all rights reserved.

The man-hour column describes how long it would take one worker to construct one unit of a task. This is a useful factor to know if you are planning on accomplishing an activity with a crew size different from the standard crew that Means suggests. The units for this column are given in man-hours, so if you are working in days, as is normal, you must divide by 8.

(Quantity)(Man-hours)/8 = Duration (in days)

See the sidebar which provides an example of duration calculations.

The example in the sidebar uses data provided by the R.S. Means Company, but it could just as well have used historical data owned by the company itself. Sophisticated companies that have developed specialty fields or are focused in particular regions and want to remain competitive need to develop their own data base. Such a data base, just like the one shown in the example, would need to be used with particular attention paid to the units, the type of work that the line number represents, the crew assumed, and any unique features pertaining to the activity being considered.

Scheduling Calculations

To this point, the project management team knows the logic of the project, the specific activities, and how long each of the activities will take. The next step is the scheduling computations. When complete, these will define how long the project will take, when each activity can specifically start and finish, and which activities can slip and for how long without affecting the completion date of the project.

The following terms are commonly used in scheduling:

Early start (ES). The early start of an activity is the earliest possible time that an activity can start based on the logic and durations identified on the network.

Early finish (EF). The early finish time for an activity is the earliest possible time that an activity can finish based on the logic and durations identified on the network. Early finish = Early start + Duration.

Late finish (LF). The late finish time for an activity is the latest possible time that an activity can finish based on the logic and durations identified on the network without extending the completion date of the project.

Late start (LS). The late start time for an activity is the latest possible time that an activity can start based on the logic and durations identified on the network without extending the completion date of the project. Late start = Late finish – Duration.

Float. The float of an activity is the additional time that an activity can use beyond its normal duration and not extend the completion date of the project. Float = Late finish - Early finish, or Float = Late start - Early start.

Sidebar

Example Duration Calculation

What is the duration in days to install 6,000 square feet of ⅜″ drywall on walls, no finish included. Assume a crew of two carpenters.

Source: Means 1994 Building Construction Cost Data
Line #: 092-608-0150

Crew: 2 carpenters
Daily output: 2,000
Man-hours: .008

Method 1

Use if the crew is the same as listed in the Means book. Assume a crew of two.

$$\text{Duration (days)} = \frac{\text{Quantity}}{\text{Daily output}}$$

$$= \frac{6{,}000 \text{ sq. ft.}}{2{,}000 \text{ sq. ft. per day}}$$

$$= 3 \text{ days}$$

Method 2

Use if the crew varies from the crew listed in Means. Assume a crew of three.

$$\text{Duration (days)} = \frac{(\text{Quantity})(\text{Man-hours})}{(8 \text{ hrs per day})(\text{Workers in crew})}$$

$$= \frac{(6{,}000 \text{ sq. ft.})(.008 \text{ hrs})}{(8)(3)}$$

$$= 2 \text{ days}$$

With a crew of two as in Method 1:

$$\text{Duration (days)} = \frac{(6{,}000 \text{ sq. ft.})(.008 \text{ hrs})}{(8 \text{ hrs per day})(2)}$$

$$= 3 \text{ days}$$

Node Notation

When using activity on arrow notation, two forms of notation are commonly used to compute the **"node times"** for the network. A node time indicates the amount of project time that must be consumed to get to this point in the schedule. Node times are also called event times.

Looking at Figure 9–3, the early time identifies the earliest possible time that the project can get to this point in the network if the logic and the durations are followed. This time is computed using the forward pass calculation (explained below) and is the early start time for all activities which originate at that node.

Figure 9–3
Two forms for computing node times.

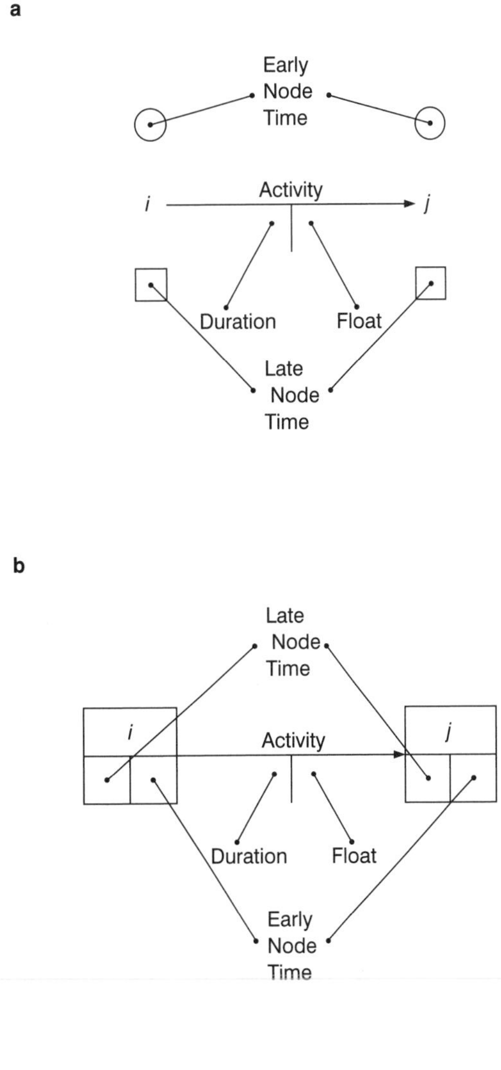

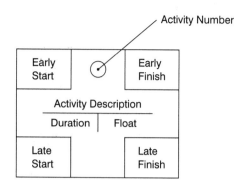

Figure 9–4
Contents of nodes in precedent notation.

The late time identified in the figure indicates the latest possible time that the project can be at this point in the network without extending the completion date of the project. This time is computed by the use of the backward pass calculation, with the late time shown being the late finish time for all activities which terminate at this node.

When using precedent notation, all activity durations and float are shown on the activity node as shown in Figure 9–4. When using this notation, the forward and backward pass calculations are done internal to the activity, with the node times equal to the activity times.

Forward Pass Calculations

In either activity on arrow notation or precedent notation the first calculation that is made is called the forward pass. It is called that because it starts at the first node and moves forward through the network. The calculation begins with the first node assigned "time 0" and the duration of each successive activity added to this time. In activity on arrow notation, the i node time of an activity with only one precedent is equal to the i node time of its precedent plus its precedent's duration. In precedent notation, the early start time of the activity with only one precedent equals the early finish time of that preceding activity (see Fig. 9–5).

In the case in which a node has more than one predecessor, the early node time in activity on arrow notation equals the largest time calculated from the converging paths; in precedent notation the early start time equals the largest early finish time of the preceding activities (see Fig. 9–6).

Illustrated in Figure 9–7 is a forward pass for a simple network diagram completed in both activity on arrow and precedent notation. Note that dummy arrows are treated the same as production arrows. Also note in the precedent diagram the use of both finish to start and start to start links and the placement of durations on the links. For instance, an SS2 means that the following activity can start 2 days after the preceding activity starts. An FS2 means that the succeeding activity can start 2 days after the preceding activity finishes.

After the early start and finish dates are established and the project completion date is found, the forward pass for the project is complete. These times identify the

Precedent Notation

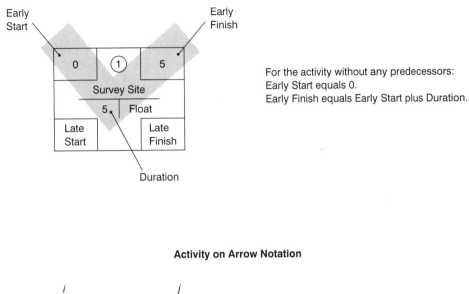

For the activity without any predecessors:
Early Start equals 0.
Early Finish equals Early Start plus Duration.

Activity on Arrow Notation

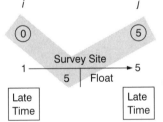

For the node without any predecessors:
The early node time equals 0.
The early event time at the j node equals the i node time plus the activity duration.

Figure 9–5
Forward pass (only one predecessor).

earliest possible time that each activity can start and finish as well as the earliest possible time that the project can be finished. In the example project, the duration for the project is 19 days. In precedent notation the early start and finish times can be read directly in the upper left and right corners of the activity box. In activity on arrow notation the early start time is the early time read at the activity's i node. The early finish time is the i node time plus the activity duration.

Example:
For Set Base Plates 2 (Activity 55–65)

Early start = Day 15

Early finish = Day 16

Precedent Notation

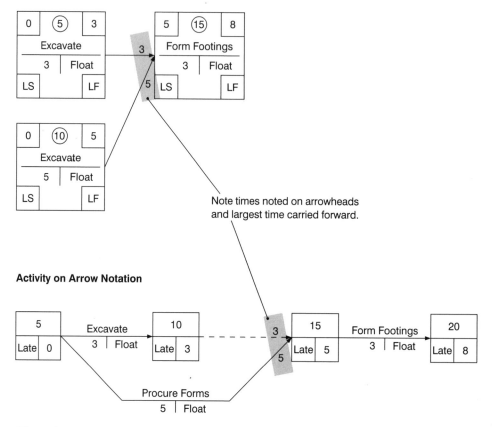

Activity on Arrow Notation

Figure 9–6
Forward pass (multiple predecessors).

It is a common error in activity on arrow notation to look to the early time at the *j* node to read the activity's early finish time. As can be seen in the previous example network, that is true sometimes, but only along the longest path through the network—which will be defined later as the critical path. Remember that the early time at the *j* node represents the earliest time that the project is expected to get to that node in the network, and that represents the early start for all activities which originate from that node.

Backward Pass Calculations

The backward pass is conducted following the completion of the forward pass. The backward pass cannot be done until the project duration is known. The backward

Activity on Arrow Notation

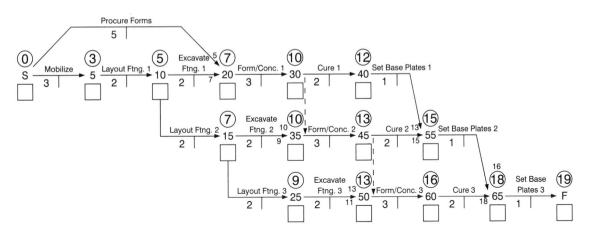

Precedent Notation

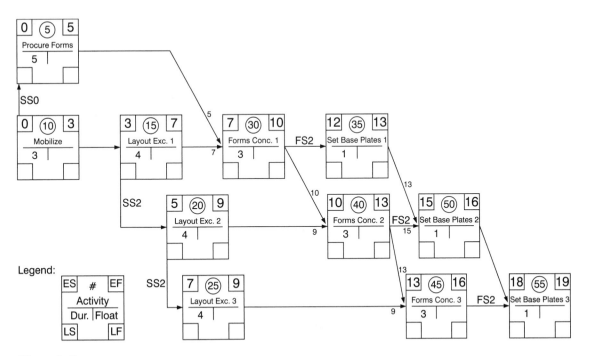

Figure 9–7
Example project: forward pass.

229

pass begins with the last node in activity on arrow notation, or the last activity in precedent notation. In activity on arrow notation the project duration is the late event time at the last node. In precedent notation the project duration is the late finish time for the last activity. The backward pass is completed by working backwards (from finish to start), subtracting the activity's duration from the previous late time. In activity on arrow notation, the late event time at the *j* node of an activity with only one successor is equal to the late event time at the *j* node of the successor minus the successor's duration. In precedent notation the late finish time of an activity with only one successor is equal to the late start time of the successor (see Fig. 9–8).

In the case where a node has multiple successor activities, the late event time for the node equals the smallest of the succeeding nodes' late times less the activi-

Precedent Notation

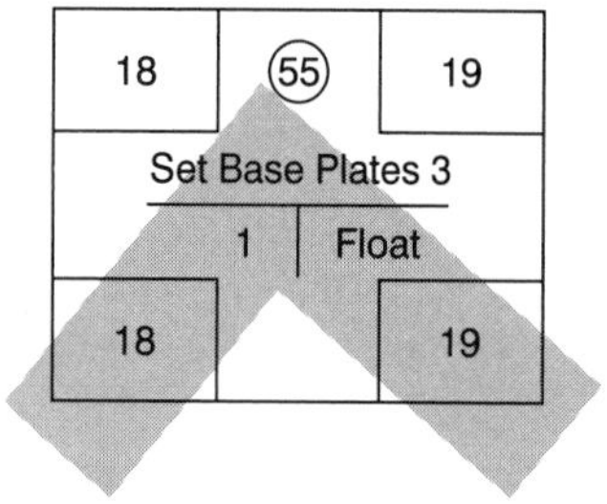

For the activity without any sucessors Late Finish equals Early Finish and the Project Duration.
Late Start equals Late Finish minus Duration.

Activity on Arrow Notation

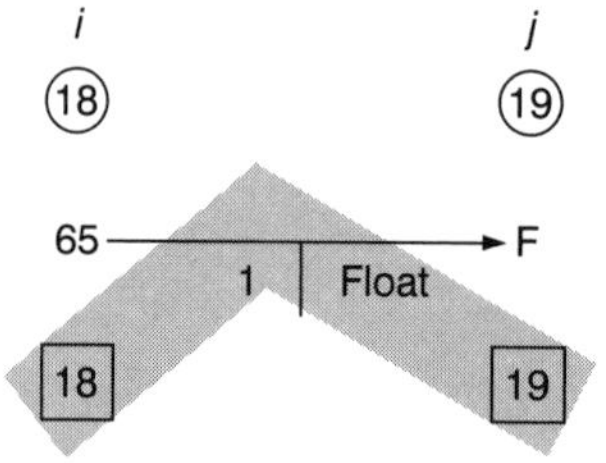

For the node without any sucessors the late node time equals the early node time and Project Duration
The late event time at the *i* node equals the *j* node time minus the activity duration.

Figure 9–8
Backward pass (only one successor)

Precedent Notation

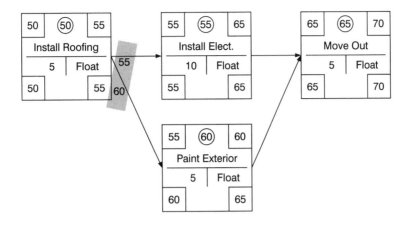

Activity on Arrow Notation

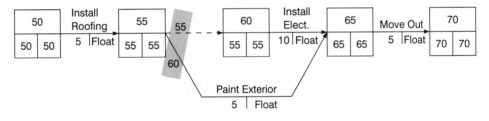

Figure 9–9
Backward pass (multiple successors).

ties' durations. In precedent notation the late finish time of an activity equals the smallest late start times of all the succeeding activities (see Fig. 9–9).

The backward pass now completed for the example network previously displayed is shown in Figure 9–10. Note the treatment of both lags and dummies.

The backward pass provides the information necessary to calculate the late start and finish times which define the latest time that an activity can start or finish and not delay the scheduled completion time for the project. In precedent notation the late start and finish times can be read directly on the activity node in the lower left and right corners of the node box. In activity on arrow notation the *j* node late time is the late finish time for the activity. The late start time is calculated by subtracting the activity's duration from the late finish time.

Activity on Arrow Notation

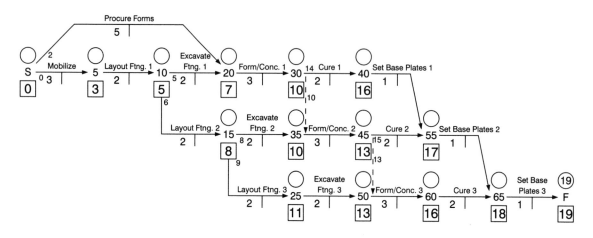

Precedent Notation

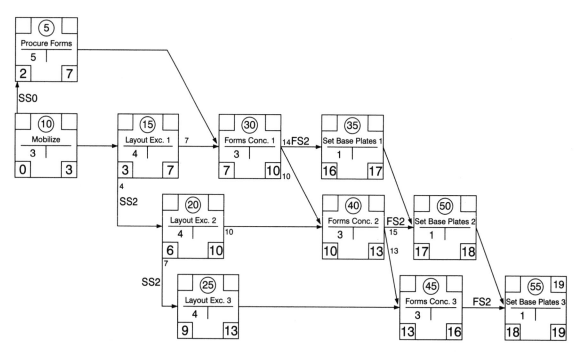

Figure 9–10
Example project: backward pass

232

Example:
For Procure Forms (Activity S-20)

> Late Finish = 7
>
> Late Start = 2

Together, the forward and backward pass calculations provide a clear picture as to the timing of each activity in the network. The forward pass defines the early time, or early window of opportunity for an activity, while the backward pass defines the late times, or late window of opportunity for the activity. The activity may be run at the early time, late time, or some combination of both. Unless the logic is wrong or the durations are defined in error, the activity cannot be done earlier or later than the times defined on the network (see Fig. 9–11).

Float Calculations

As can be seen in Figure 9–11, the activity can be done either early or late, and the difference between the late start time and early start time is equal to the difference between the late finish time and the early finish time. This difference is called **float** time. Float, also called slack, is extra time that occurs in some networks. There are three formulas that can be used to calculate float time:

> Float = LS – ES
>
> Float = LF – EF
>
> Float = LF – (ES + Duration)

Figure 9–11
Activity times in perspective.

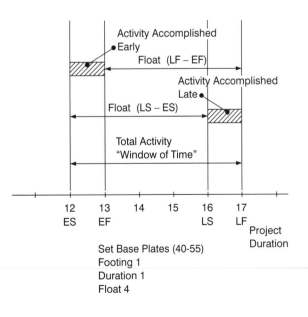

Examining Figure 9–12, you will see there are nine paths through the network:

Paths:
   1. S-5-10-20-30-40-55-65-F
   2. S-20-30-40-55-65-F
   3. S-20-30-35-45-55-65-F
   4. S-20-30-35-45-50-60-65-F
   5. S-5-10-20-30-35-45-55-65-F
   6. S-5-10-20-30-35-45-50-60-65-F
   7. S-5-10-15-35-45-55-65-F
   8. S-5-10-15-35-45-50-60-65-F
   9. S-5-10-15-25-50-60-65-F

If you were to calculate the durations of each of the paths they would be as follows:

Durations:
   Path 1 = 15
   Path 2 = 13
   Path 3 = 15
   Path 4 = 17
   Path 5 = 17
   Path 6 = 19
   Path 7 = 16
   Path 8 = 18
   Path 9 = 17

Examine the longest path, Path 6. This path has no float, indicating that any delay in any of the activities along this path will delay the completion of the project. Also notice that the float for each activity in this path equals 0. This is called the **critical path** because any delay in this path will cause the schedule to be delayed. The next section explains this further. Notice Path 2: this path shows a duration of 13 days. Further examination shows that two activities in this path have 0 float and two activities in the path have 4 days float. The two "0-float" activities can be explained because these activities are also part of Path 6, the longest path in the network. What about activities 30–40 and 40–55? Can both activities use the 4 days of float that has been assigned? The answer is no. If activity 30–40 uses the 4 days of float activity, then 40-55 becomes critical. The float that was defined above is called **total float**, which defines the number of extra days that can be assigned to a particular activity. Any of the activities in the network that have float can use this extra time, but not all of them. To use float often makes other activities critical, such that if they also take additional time the project will be delayed. In the case of activities

Activity on Arrow Notation

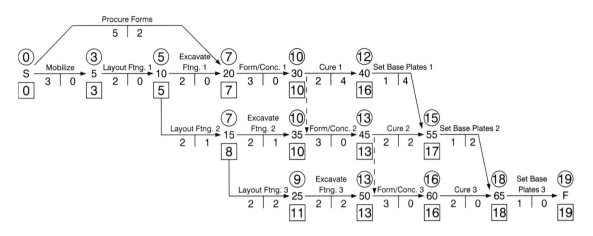

Precedent Notation

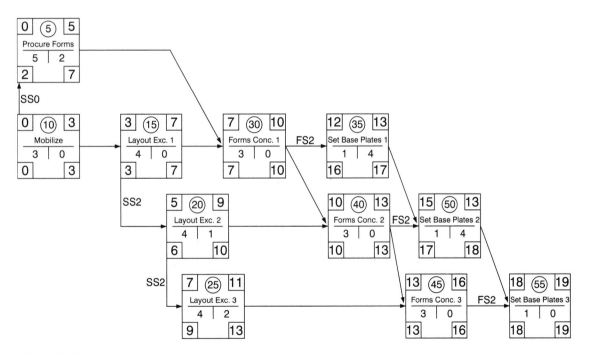

Figure 9–12
Example project: total float calculations.

Activity on Arrow Notation

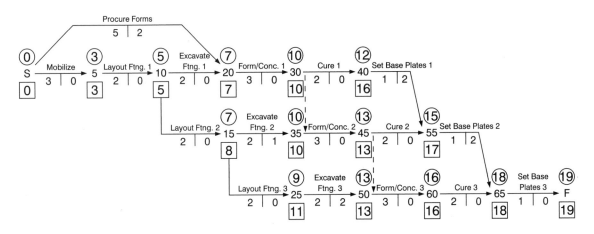

Precedent Notation

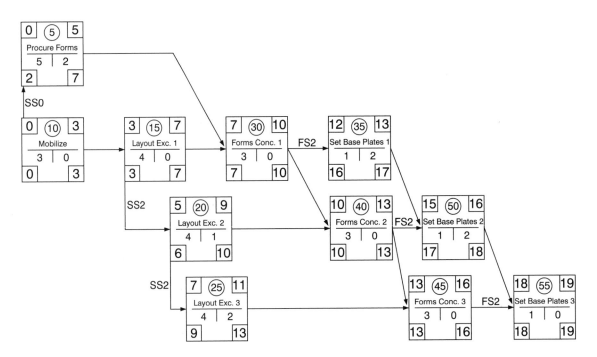

Figure 9–13
Example project: free float calculations.

30–40 and 40–55, if both activities were to use the additional 4 days, then Path 2 would have a duration of 21 days, delaying the project.

Free float is another type of float that can be included in a network (see Fig. 9–13). The formula for calculating free float is as follows:

Free float = ES (of succeeding activity) – EF (of activity in question)

Free float is useful to know because it defines the amount of time that an activity can be delayed without taking float away from any other activity. In the case of activities 30–40 and 40–55 in Figure 9–13, 30–40 would show 0 days of free float since if it uses any float it will be delaying the early start of activity 40–55 and thereby utilizing activity 40–55's float. Activity 40–55 shows 2 days of free float because it can be delayed 2 days without impacting any activity, but after 2 days it affects the early start of activity 55–65. With free float available for an activity, a project manager knows that the float can be used without making other activities in the network critical. Another example of free float in the example project is activity S–20. This activity shows 2 days of free float. The activity can be completed on day 7 without affecting any other activity on the project. In the case of a string of activities all of which show the same number of days of total float, the last activity in the string will receive the free float with the other activities showing 0 free float. That is because only this activity can utilize the float without taking the float away from other activities.

It is best to think of float as flexibility. By utilizing float the project team has the ability to shift activities around to more efficiently utilize people, equipment, and space. This flexibility is worth money to the project, and for that reason most good project managers do not want to give away any float. This topic will be covered in the control section of the book, Section Four.

Critical Path

As was identified above, Path 6 is the longest path through the network. It has 0 float and by definition is called the critical path. Any delay along this path delays the project (see Fig. 9–14). Note that there can be more than one critical path in a project. It is important for the project manager to identify these paths and examine them closely because control of the time on these paths defines the control of the project. If the duration of the project is too long, the activities that must be shortened lie on the critical path and must either be planned differently, accelerated, or eliminated.

If the critical path is shortened, more critical paths get created and the probability of the project being delayed due to some unplanned or unexpected happening increases. It is important for management to study both the critical and the almost critical paths closely to try to anticipate and plan for all possible disasters that can occur. The fact that these unforseen disasters do happen is why float is so important.

Activity on Arrow Notation

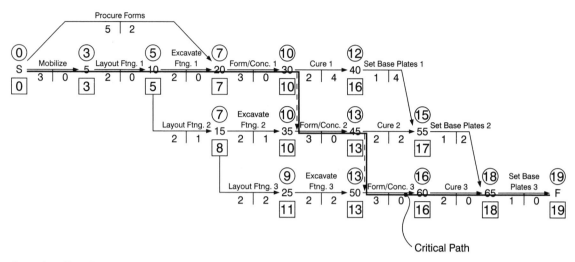

Precedent Notation

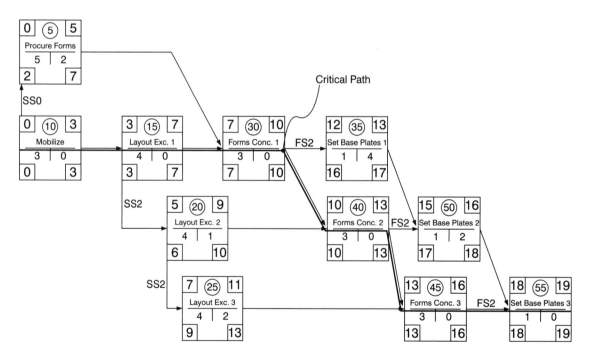

Figure 9–14
Example project: critical path.

Float in a project is a built-in contingency that provides the project team with the ability to respond to the many unplanned occurrences in a project.

Conclusion

It is important to remember that scheduling is a team process that involves the input and support of all the key project players. The process begins with the identification of the project goal and ends with the computation of project duration and activity start and finish times and float. The process is iterative—rarely does the first cut at the schedule result in an output acceptable to everyone. The main reason for this is that the interconnections between the equipment, work force, calendar, and site are not completely known until the scheduling process is complete.

A schedule is a communication tool that helps everyone better understand the project at hand. Even if the schedule is never used again (not an ideal situation), the process of developing the schedule would still have provided major benefits to the project team. With a schedule and estimate in hand, the project team is ready to apply these basic tools to better plan and control the construction project.

Chapter Review Questions

1. A forward pass is used to determine late start and late finish times.
   __ T   __ F

2. The time for completing a project is equal to the sum of the individual activity times.
   __ T   __ F

3. Total float equals:
   a. Late finish time minus early finish time
   b. Late start time minus early start time
   c. Late finish time minus (early start + duration)
   d. All of the above

4. The maximum amount of time that an activity can be delayed without extending the completion time of the overall project is called:
   a. Duration
   b. Float
   c. Critical path
   d. None of the above

5. Project float is most useful in balancing project resources.
   __ T   __ F

Questions 6–11 refer to the accompanying figure.

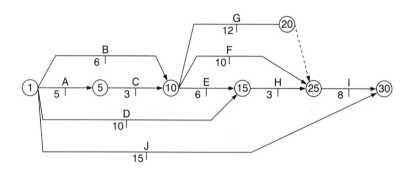

6. The ES time for D is _____.

7. The LS time for I is _____.

8. The LF time for E is _____.

9. The EF time for J is _____.

10. Activity H has _____ days of total float.

11. Activity D has _____ days of free float.

Exercises

1. Explain the best way to determine the duration of a production activity. What is the best way for a procurement activity? What for an administrative activity?

2. Obtain the contract documents for a moderate size project. Prepare a CPM network, compute durations, perform a forward and backward pass, and prepare a table which lists activities, durations, ES, EF, LS, LF, and TF. Work this problem using both activity on arrow and precedent notation.

Sources of Additional Information

Callahan, Michael T., Daniel G. Quackenbush, and James E. Rowlings. *Construction Project Scheduling*. Englewood Cliffs, NJ: Prentice Hall, 1992.

Horsley, William F. *Means Scheduling Manual*, 3d ed. Kingston, MA: R. S. Means Co., Inc., 1990.

Iannone, Anthony L., and Andrew M. Civitello, Jr. *Construction Scheduling Simplified*. Englewood Cliffs, NJ: Prentice Hall, 1985.

Naylor, Henry. *Construction Project Management: Planning and Scheduling*. Albany, NY: Delmar Publishers, 1995.

O'Brien, James J. *CPM in Construction Management*, 4th ed. New York: McGraw-Hill, 1993.

Pierce, David R. *Project Planning & Control for Construction*. Kingston, MA.: R.S. Means Company Inc., 1988.

Popescu, Calin M., and Chotchai Charoenngam. *Project Planning, Scheduling, and Control in Construction: An Encyclopedia of Terms and Applications*. New York: John Wiley & Sons, Inc., 1995.

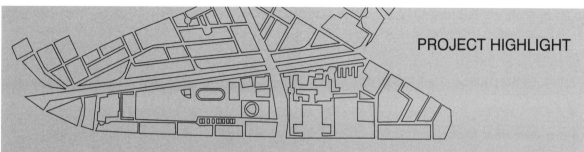
SECTION THREE

Scheduling

Nancy E. Joyce

Before a project schedule could be developed for Building 16 and 56 there were certain owner decisions that needed to be made. How the project would be built was of prime importance and was examined by the team early in the process. One of the owner's stated goals was that researchers in the buildings would only be moved once. However, as we examined the issues surrounding this, it became apparent that this would be difficult to achieve.

Moving a laboratory space is far more disruptive than moving an administrative or teaching space. Timing is important if the researcher has a long term experiment going on, and the amount of downtime can be excessive if equipment needs to be recalibrated. Continuation of research grant money hinged on timely results, and young researchers were also concerned about tenure. Therefore, minimizing the number of moves was important. However, these researchers were on the lower floors of both buildings, which meant that construction would have to occur above them and below them. Plumbing, electrical, and mechanical shafts would be cut into their spaces, and new sprinkler work would have to be put in place.

As part of our analysis, we looked at the schedule and cost implications associated with pursuing several options. A one-phase scheme assumed we could renovate both Building 16 and 56 with the researchers in place. With this, however, only minimum renovation work to the areas they were occupying would have been possible. The two-phase scheme looked at renovating all of Building 56 by using Building 16 as temporary space and then moving all of Building 16 into Building 56 permanently. This, of course, meant that some researchers would be moving twice. It also limited the possible adjacencies for the various research groups. With a three-phase scheme, which is the one the Institute initially was hoping we could accomplish, we would have renovated the upper floors of Building 56, moved the researchers up, renovated the lower floors of 56, moved researchers over from Building 16, and then renovated Building 16 in its entirety. With the probability of unanticipated power outages, dust, vibration, noise, and safety issues, this particular option was not optimal.

From a cost and schedule point of view, the analysis clearly showed that the two-phase scheme was the least expensive and least disruptive for the Institute. Once a realistic plan was developed for moving the researchers temporarily, the Institute felt comfortable going forward with the two-phase renovation of the buildings.

After the owner decided to construct the project in two phases, there still remained the question of how to phase the design and bidding of the project. The architect's recommendation was to design and bid both buildings at once. From the architect's perspective, once the team was assembled for the project, it was most efficient to complete all the design work in one phase. They faced the probability that some of their team would be reassigned if there was a lag between phases and that the knowledge gained would be lost if new personnel were assigned to a second phase. Because the MEP systems in Phase One also would serve Phase Two, there was concern that design of these would be compromised if the design team didn't understand the requirements for both phases. Also, if the design was split and the project was bid separately by phase, then there would be the potential of having two different prime subcontractors on site. Since the mechanical, electrical, and plumbing systems were being shared by both buildings, the work pertaining to these systems was interdependent.

From the owner's point of view, however, there was a good probability that the program could change before the second phase was under construction. Researchers moved in and out of the Institute; new grants dictated new spaces and some of the Phase Two occupants might have been relocated elsewhere before occupancy. Also, early schedules showed a distinct advantage in splitting the design into two phases (a savings of 6 months in design time), thereby allowing the project to be finished 6 months earlier. To address the architect's concerns the decision was made to carry both phases through design development. This would bring the design to a point where all the needs were known, but would delay the formal detailing of Phase Two until Phase One was completed. Also, to avoid the overlap of two different subcontractors, pricing would be obtained from the subcontractors at bid time for Phase One with an option for the second phase.

By choosing a phased delivery, the owner got the combination that was needed to complete the job in the most cost effective way. The schedule could be accelerated, the costs could be kept down by negotiating both phases, and any changes to the second phase of the project would have a minimum schedule and cost impact.

Once the decisions were made about the design and construction phasing, the project team was able to put together a project schedule. The architects established key milestones for the design aspects of the project; the owner added Institute calendar concerns; and the construction manager put together the construction durations. All the members of the project team came together to formulate the project schedule. To facilitate the process, the team began by putting the calendar dates by month on a large board and using "yellow stickies" to manipulate design and construction activities. This was a whole-day session and involved a lot of discussion between team members. At the conclusion of the meeting, the construction manager incorporated all the information into an overall project schedule, which was sorted by responsibility so that all owner activities, architect's activities, and builder activities could easily be accessed for quick reference (see Figure A).

To minimize the construction duration, the schedule called out early packages to be developed by the architects and bid by the construction manager before the completion of the overall documents (see Figure B). These included long-lead equipment such as fume hoods, laboratory casework, electrical generator, and fire alarm panels and long-lead subcontract packages such as curtainwall, elevator, and demolition. The lead times for the equipment were figured backwards, starting from the date when the items would be needed physically on the job. The lead times for the packages were also figured backwards, but the required completion date was used as the starting point. When the full packages were bid out, the long-lead equipment would become the responsibility of the appropriate subcontractor.

Formulation of the schedule was a process that took the collaboration of the whole team and had to be reviewed constantly and consistently to ensure that all activities were on target and that all team members understood the relationship of their specific activities to other project activities. If changes occurred to any one activity, it could affect other activities. By keeping the schedule updated and highlighting the interconnections, the schedule served as a communication tool for the project team. It was the one place where the entire project was mapped out with activities for all members of the team to use as a guide.

Description	Original uration	Early Start	Early Finish
BCC			
Schematic Cost Estimate Impact	1d	23JAN95 *	23JAN95
Hazardous - Bid/Award	35d	27FEB95 *	14APR95
Site Investigation of Utilities	10d	06MAR95 *	17MAR95
Mock-up Exterior Wall	15d	13MAR95 *	31MAR95
Elevators Bid/Award	28d	29MAR95 *	05MAY95
Hazardous Remove	40d	04APR95 *	29MAY95
Long Lead Bid/Award (Item 10)	7d	21APR95 *	01MAY95
Elevator Shop Drawing	12d	15MAY95 *	30MAY95
Cost Estimate	30d	29MAY95 *	07JUL95
Inventory Lab Casework	92d	29MAY95 *	03OCT95
Elevator Manufacturing	69d	30MAY95 *	01SEP95
Bid/Award - Enabling Doc	20d	06JUN95 *	03JUL95
Asbestos - Bid/Award	12d	29JUN95 *	14JUL95
Bid/Award Biemann Space	20d	14JUL95 *	10AUG95
Enabling Work	30d	14JUL95 *	24AUG95
Notification Asbestos Removal	11d	24JUL95 *	07AUG95
Asbestos Removal	34d	07AUG95 *	21SEP95
Biemann Space Construction	40d	22AUG95 *	16OCT95
Elevator Work (Building 16)	86d	04SEP95 *	01JAN96
Laboratory Mockup	25d	05SEP95 *	09OCT95
Bid/Award Demolition	25d	11SEP95 *	13OCT95
Demolition Work	30d	16OCT95 *	24NOV95
Long Lead Bid/Award (Item 6)	20d	23OCT95 *	17NOV95
Long Lead Bid/Award (Items 1 - 5)	20d	01NOV95 *	28NOV95
Long Lead Bid/Award (Items 7 - 9)	20d	04DEC95 *	29DEC95
Facade 16 & 56 Bid/Award	15d	25DEC95 *	12JAN96
Building 56 Construction	217d	01FEB96 *	29NOV96
Mobilize - Construction Start (Building	1d	10JUL96 *	10JUL96
Bid/Award - Building 16	45d	01OCT96 *	02DEC96
Construction - Building 16	238d	02JAN97 *	01DEC97
Construction - Basement 56 and	66d	15DEC97 *	16MAR98

Figure A
Project schedule sorted by responsible team members.

EAI

Activity	Dur	Start	Finish
Schematic Design	15d	02JAN95	20JAN95
Schematic Drawings Re-issue	10d	09JAN95	20JAN95
User Drawings - User Groups 1 & 2	30d	09JAN95	17FEB95
Exterior and Public Space Design	20d	16JAN95	10FEB95
Exterior Wall Consultant	18d	23JAN95	15FEB95
Design Development	90d	23JAN95	26MAY95
Meetings, Round 1 - User Groups 1 &	20d	06FEB95	03MAR95
Drawing Revisions - User Groups 1 & 2	50d	13FEB95	21APR95
Code Review Meeting	0	16FEB95 *	
Exterior and Public Space Design	15d	20FEB95	10MAR95
Out-of-Phase Users, Schematic	20d	20FEB95	17MAR95
Meetings, Round 2/Signoff - User	25d	27FEB95	31MAR95
Exterior and Public Space Design	20d	20MAR95	14APR95
Out-of-Phase Users, Design	60d	20MAR95	09JUN95
Engineering Drawings	41d	27MAR95	22MAY95
Develop Demolition Document	44d	16MAY95	14JUL95
Contract Documents (Building 56)	120d	10JUL95	05JAN96
Confirm Design - Building 16	68d	27NOV95	28FEB96
Contract Documents - Building 16	152d	01MAR96	30SEP96

MIT

Activity	Dur	Start	Finish
MIT Salvage	95d	06MAR95	14JUL95
MIT Review and Approval Period	20d	07JUL95 *	03AUG95
Review/Approve Demolition Document	6d	14JUL95 *	21JUL95
Approval - Budget	1d	28JUL95	28JUL95
City Permit Review (60% Doc)	1d	26SEP95 *	26SEP95
Building Permit	1d	23NOV95 *	23NOV95
Occupancy - Building 56	20d	01NOV96 *	28NOV96

TEAM

Activity	Dur	Start
MIT Design Development Cost Review	0	07JUL95 *
MIT Review Meeting	0	17JUL95 *
MIT Review Meeting	0	31JUL95 *
Design Review Meeting	0	14AUG95 *
MIT Review Meeting	0	28AUG95 *
MIT Review Meeting	0	11SEP95 *
MIT Review Meeting	0	18SEP95 *
MIT Review Meeting	0	25SEP95 *
MIT Review Meeting	0	02OCT95 *

Massachusetts Inst. of Technology
Project Schedule
Building 16 and 56 Renovation

Beacon CONSTRUCTION

Start date 02JAN95
Finish date 16MAR98
Data date 02JAN95
Run date 14NOV95
Project Name MIT030/
Project type SureTrak
© Primavera Systems, Inc.

Massachusetts Institute of Technology
Building 16 and 56 Renovations
Early Packages - Lead Times

LONG LEAD EQUIPMENT

ITEM	PACKAGE COMPLETE	OWNER APPROVAL	BID / PURCHASE	SHOP DRAWINGS	SHOP DWGS REVIEW	DELIVERY AFTER APPROVED SHP DWGS	NEEDED ON JOB
1. Fume hoods	01-Oct-95	2 weeks	3 weeks	3 weeks	4 weeks	16 - 18 weeks	June 96
2. Laboratory casework	01-Oct-95	2 weeks	3 weeks	3 weeks	4 weeks	16 - 18 weeks	June 96
3. Electrical substation	01-Oct-95	2 weeks	3 weeks	3 weeks	4 weeks	18 weeks	June 96
4. Electrical generator	01-Oct-95	2 weeks	3 weeks	3 weeks	4 weeks	16 weeks	June 96
5. Fire alarm panels	01-Oct-95	2 weeks	3 weeks	3 weeks	4 weeks	18 weeks	June 96
6. Air handlers (built-ups)	15-Sep-95	2 weeks	3 weeks	3 weeks	4 weeks	16 - 20 weeks	June 96
7. Air compressors	01-Nov-95	2 weeks	3 weeks	3 weeks	4 weeks	12 weeks	June 96
8. Vacuum pumps	01-Nov-95	2 weeks	3 weeks	3 weeks	4 weeks	12 weeks	June 96
9. Pure water system	01-Nov-95	2 weeks	3 weeks	3 weeks	4 weeks	12 weeks	June 96
10. Biemann air handler	15-Mar-95	1 week	3 weeks	3 weeks	4 weeks	10 weeks	Aug 95

LONG LEAD PACKAGE

	PACKAGE COMPLETE	OWNER APPROVAL	BID / PURCHASE	SHOP DRAWINGS	SHOP DWGS REVIEW	CONSTRUCTION DURATION	COMPLETION
1. Facade pacakge (both bldgs)	Dec 95	2 weeks	4 weeks	4 weeks	4 weeks	n/a	n/a
2. Elevator (both bldgs)	Mar 95	2 weeks	3 weeks	2 weeks	2 weeks	26 weeks	Nov 95
3. Demolition	Sept 95	1 week	3 weeks	n/a	n/a	8 - 12 weeks	Jan 96
4. Temporary services	June 95	1 week	n/a	3 weeks	2 weeks	n/a	Oct 96
5. Biemann package	June 95	1 week	n/a	n/a	n/a	n/a	Oct 96

Figure B
Early packages and equipment were laid out in a spreadsheet format so that the team could analyze the dates and make adjustments if necessary.

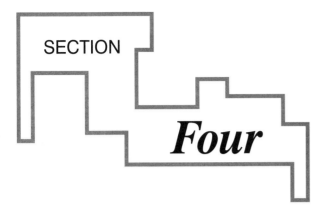

Project Control

This section will look at the use of project estimating and scheduling information in the control of the construction project. The estimate and schedule are prepared upfront, but as the project evolves, changes occur and the project team must work to maintain control of the cost and time budgeted for the project.

Chapter 10 will first look at the basic project control model. It will explain how the project team can use actual project information to measure how well the project is proceeding. The chapter will next explain how the estimate and schedule can be "integrated" to arrive at the optimum duration for the project. Chapter 11 will look at how resources can be attached to a schedule to project cash and equipment needs, as well as how float can be used to level resource consumption. Chapter 12 will introduce cost accounts, progress measurement, updating, and project documentation. It will discuss as well as provide examples of project reports.

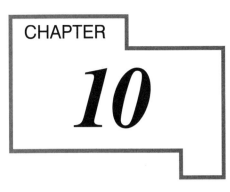

Fundamentals of Project Control

From studying this chapter you will learn:

1. The objectives of a project control system
2. How to diagram the project control cycle
3. How to calculate the indirect costs for a project
4. How to calculate the optimum duration for a project

Introduction

The project management cycle begins with the identification of the owner's project objectives. The owner, designer, and construction professional together design, estimate, and schedule the project in accordance with the owner's objectives. The larger and more complicated the project, the more coordination is required to manage this process and ensure that the owner's requirements are met.

Project control is a process. Its purpose is to guarantee that the design requirements, budget, and schedule are met by the project team. If any project objective begins to slip, the project control system should identify this deviation early and allow a correction to be made.

Project control begins with a plan that identifies the objectives of the project, with specific checkpoints throughout the project cycle. Envision the plan as a roadmap which allows the project team to constantly monitor and make corrections as necessary. A project plan is generally composed of design documents that establish the quality objectives, an estimate that establishes the budget, and a schedule that establishes the timing for the project.

Project control is an action based process that requires the continual monitoring of the project's operations. To guarantee success it is important that the actual work, its cost, and its duration be documented and compared to the original work plan. Any deviations must be noted and adjusted for. Actual durations and costs should be noted and used for future estimates and schedules.

Forecasting is also critical to the project control process. The expected cost to complete the work and the expected time of completion must be continually updated and reported. The control process must document progress and allow the project team to adjust to occurrences that were not expected, such as change orders, strikes, or bad weather. A control system includes a reporting system to notify all necessary parties of the project status. This allows the input of outside technical experts and senior managers to assist and plan adjustments as necessary. Lastly, the project control system is iterative: the process occurs over and over, encouraging a continual adjustment to the plan of the project team.

Basic Control Theory

Project Control Objectives

An effective project control system is essential to the successful delivery of a construction project. Projects of substantial size or complexity need to be continually managed to guarantee any possibility of success. These projects require good coordination between the involved disciplines, utilizing budget and schedule milestones, monitoring of actual progress, reports, and adjustment as necessary.

A project control system begins with the establishment of project standards. Control standards are to the project team as lap times are to an Indy racer (see Fig. 10–1).

The project team uses control standards to continually check progress against acceptable standards. Just as performance data for a single lap can be projected to overall race performance, project performance at key milestone dates can be projected to successful completion at the end of the project. Standards for quality control would be defined by the drawings and specifications. Drawings define the quantity of work required, locations, and widths and heights. Specifications define the quality of work, defining performance standards, addressing issues such as alignment, compression strengths, and finishes. The project estimate establishes the over-

Figure 10–1
Control standards are like lap times: performance on a single lap can be projected to predict overall performance.
Photo by Paul C. Betts

all budget for the project, as well as milestone costs for specific phases of work. The schedule defines when specific work items need to be accomplished. Estimate data can be integrated with schedule information to provide additional project standards.

The measurement standards just defined serve as the goals for the project team. They are targets, which if well thought out can serve to organize a very complicated project. Each project participant can be given a task, a budget, and a time frame in which to accomplish the job. As an example, a carpenter can be given the task of framing 100 feet of interior partition, a budget of $1,500 for materials, and a duration of 2 days.

The second component of the project control system is measurement of actual performance on the project. Actual performance compared to planned performance will provide the management team with feedback as to how well the project is proceeding. Also, figures on actual performance can be kept for future reference when estimating and scheduling similar projects.

Measuring actual performance is a complicated process since so much information needs to be gathered from so many different sources. An on-site review of daily reports can provide actual durations for work activities and the percent complete for the project. Worker time sheets and equipment logs will define how many hours were spent working on a particular task. Purchase orders, delivery tickets, and receipts will provide actual costs for material, equipment, and vendor purchases. Weather conditions and unforseen circumstances can be picked up from daily reports (see Fig. 10–2).

The information from these reports makes up the raw material for a project control report which can be organized to inform project participants. As an example, a sitework superintendent may have been assigned a site clearing task as detailed in Figure 10–3.

The processing of the data and the production of the report are critical to providing the user with feedback as to how the work is proceeding. In the example illustrated, the hours expended are greater than what was budgeted for the work. The inconsistency can be explained through an analysis of several possible options:

1. Poor worker performance
2. Incorrect reporting
3. Differing site conditions
4. Wrong equipment assigned
5. Bad weather
6. An error in the estimate

The next step for the management team is to determine which of the above reason(s) explains the **variance** (difference) noted in the report. A check of past daily reports should verify the weather for the reporting period. A walk around the site with the superintendent and estimator would allow a check on the approach being used, the actual site conditions, and the time and quantities assigned in the estimate. If the

Figure 10–2
Daily reports are one source of information on actual project performance, providing information as to weather conditions, work accomplished, visitors, and key deliveries.
Courtesy of Walsh Brothers, Inc.
Photo by Don Farrell

	Budget	**Actual**	**Variance**
Quantity (Acres)	5 Acres	5 Acres	
Equipment (Hrs)	40 Hrs	48 Hrs	(8)
Labor (Hrs)	40 Hrs	48 Hrs	(8)

100% Complete

The Budget column reflects the estimated and scheduled standards for the site clearing activity.

The Actual column indicates the equipment and labor hours expended to complete the job.

The Variance column notes any difference between the Actual and Budget.

This job required one worker and one dozer for 6 days instead of the budgeted 5.

Figure 10–3
Site clearing report.

estimate, site conditions, and weather all appear to be consistent with expectations, then worker performance should be examined. Worker inexperience, hours worked versus reported, or supervision problems could all explain the performance of the task.

The key to the reporting process is that the inconsistency be identified quickly and the proper adjustment made in a timely manner. Good reports are accurate, show variances between the budgeted amounts and the actual amounts, forecast goals for future work periods, and suggest corrective actions. Timeliness is also important, since if all of the above is accomplished, but done too late for any corrective action, the control effort has been wasted.

Project Control Cycle

The process of project control can best be illustrated as a control loop that repeats itself in a similar pattern. The project control cycle is illustrated in Figure 10–4. The control process begins with the initial project plan (item 1). The project plan includes a budget, schedule, and other planning information such as staffing and administrative procedures. This project plan is the result of the work of many different people, requiring technical input and a commitment from management. The estimate and schedule are the primary control tools described here, but the level of quality as defined by the design documents would also be managed during construction.

The project plan would be used to initiate the field operations, as shown in the figure (item 2). The order in which the work is completed and the type and level of staffing would be determined by the project plan. The initial plan would identify the resources such as equipment, people, and materials that are needed at the job site. The field supervisors are responsible for the productive utilization of these resources. As was mentioned in the scheduling section of this book, the field people are an important resource in determining the durations and resources needed to accomplish the construction activities.

The lightning bolt (item 3) represents the impact of external factors such as labor strikes, vandalism, bad weather, or other events that are difficult to predict and affect the field operations. Estimates and schedules generally provide for some inefficiencies, but excessive impacts can severely disrupt the field activities and thus the schedule. The arrow between field operations and the cost/schedule block represents the processing of actual information from the field. Schedule information such as activities completed or partially complete, deliveries, or submittals received and approved or disapproved need to be recorded. Cost information such as equipment hours, labor hours, material purchases, or subcontracts signed also need to be recorded. The process involves many people in the organization and therefore, for it to work, it takes good coordination. Preprinted forms and a cost-coding system facilitate this effort. A good cost-coding system will ensure that the data is assigned correctly and thereby allows a true comparison of actual performance versus planned.

The cost/schedule engineers block (item 4) represents the coordination of data from the field and the comparison with the initial plan. The technical people who

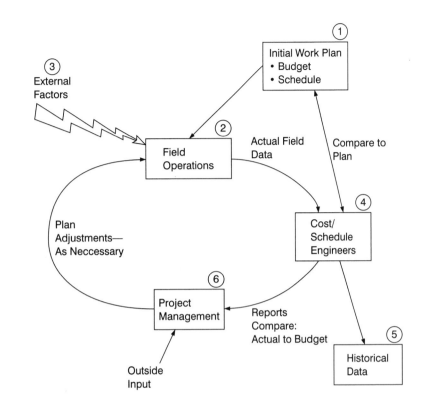

Figure 10–4
Project control cycle.

established the initial work plan are responsible for recommending adjustments to the plan based on their analysis of actual field operations. There are many reasons for a plan needing to be adjusted, but the goal is to keep the budget and schedule on target. The best way to accomplish this is for the technical people to make timely and accurate recommendations to the project manager for a final decision.

Historical data (item 5) represents the permanent storage of information for use by the company in future job planning. The actual field information that is collected represents the type of performance that is being delivered and should be used to update the company data base. This is how a data base is built and maintained. The cost-coding system that the company establishes as well as the accuracy of the data capture and processing dictates how useful the data base will be.

The arrow between the cost/schedule engineers and the project management blocks (items 4 and 6) represents the dissemination of status reports to the project team. For the reporting process to be useful, it must deliver accurate information to the right people on time. The reports should be sorted in a way that managers get only the information that pertains to their job. The reports should also be sorted to indicate the more important activities first with key variances noted.

Project management (item 6 in figure) represents the final decision point in the control process. The goal of a project control system is to deliver to the decision makers accurate and timely project status information so that intelligent decisions can be made. A plan has been set and actual progress has been measured. Management must now decide the best course of action to take. The diagram also shows outside input support, which could come in the form of technical staff or consultant support.

The arrow from project management back to field operations represents the completion of the project control system. Management has made a decision and final instructions are now being given to the field. Adjustments may be made in the project plan, or the instructions may be to continue on as originally scheduled. For these instructions to be effective, they must be delivered soon enough to be smoothly implemented.

The project control cycle is a feedback loop providing all project participants with a measure as to the success of their past decisions. An estimator can see the accuracy of the estimate made, just as a project superintendent can see if productivity was increased in the last reporting period. This loop allows learning to occur and adjustments to be made. Without a project control cycle, project people might continue to make the same mistakes and would have little opportunity to measure the effect of specific decisions.

The project control cycle can be repeated as often as is necessary to control a particular project—monthly, weekly, daily, or even hourly for tightly controlled maintenance operations. However, the more frequent the reporting process, the more cost incurred by the company because of the additional time necessary to collect, process, and interpret results.

Optimum Project Duration

In preparing a project for construction, the project team has broken the project down into activities and put together a schedule and an estimate. In preparing these, the team assumed normal conditions for each activity, providing a baseline estimate and schedule for the job. The next step is to look for opportunities to adjust and streamline the project activities.

Crashing

Crashing a project is the term used to describe the process of accelerating an activity or multiple activities to shorten the overall duration of a project. By adding additional people or equipment or by working additional hours, an activity's duration can be shortened, and if it is a critical activity, this will shorten the project as well. Activities are crashed for different reasons:

1. An activity may need to be completed by a specific date for contractual reasons.

2. Some activities can be accomplished more economically during a certain time of the year, encouraging the acceleration of activities.

3. The cost to accelerate an activity that shortens the project's duration may be less expensive than the cost of running the project for the same period.

When an activity is crashed, the direct costs for that activity increase. Direct costs are the costs of materials, labor, and equipment directly associated with the installation or construction of the project. Crashing causes the direct cost of the project to increase because of the inefficiencies caused by accelerating the work at a rate faster than normal. People may end up working in tighter quarters, or equipment may end up sitting idle, as was described in the Santa Monica freeway reconstruction sidebar. But, as that sidebar illustrated, the increase in direct cost expenditures may be justified if indirect costs are saved or a bonus is provided.

An accelerated project can earn additional bonus money, can prevent the payment of fines or damages to the owner, or can save the company additional indirect costs. The amount of bonus to be received or fine to be paid should be described in the contract. The company's indirect costs are the costs to the company to support the project in general overhead and supervision (see sidebar, Indirect Cost Calculation).

Graphically, the relationship between these costs—indirect costs, bonuses or fines, and direct costs—can be seen in Figure 10–5. The combination of indirect costs, fines, and/or bonuses can be treated in a linear fashion on a cost-per-day basis. For every day that a project is shortened, the company will gain $1,000.

The direct cost curve shows a cost of $101,245 when the project is scheduled to be accomplished in 37 days, its normal duration. As the project is crashed, the direct costs for the project increase up to the shortest possible duration of 29 days, with a corresponding direct cost of $106,375.

The total cost curve is the combination of the indirect curve and the direct cost curve. When the project is run at its normal duration, the total cost for the project is $119,745. The project's maximum crash cost is $116,875, and, as noted by the low point on the graph, the project's optimum duration is 30 days, at a cost of $116,465.

The calculation of the indirect cost curve is accomplished as described in the sidebar, whereas the direct cost curve is derived as follows. Shortening the project by crashing one or more activities requires that at least one critical activity be accelerated. Activities that shorten the duration for the least cost are the logical first choice. If the cost of accelerating these activities is less than the indirect cost curve's cost per day, the acceleration makes sense. As can be seen in the direct cost curve in Figure 10–5, the crash cost gets progressively more expensive since as the project is shortened float is absorbed and more and more critical paths are formed. Eventually, the direct costs to shorten a project begin to exceed the indirect costs saved, and the optimum project duration is reached. The following example illustrates in detail how the process of identifying the optimum duration for the project above was calculated.

Sidebar

Indirect Cost Calculation

The indirect costs on a project are those costs necessary to keep the company in business and support the physical construction of the project. Indirect costs can be divided into project overhead, also called general conditions, and home office overhead.

Project overhead can be further broken down into the following categories:

Organization and personnel: Superintendent, field engineer, watchman, and other people dedicated to the safe and efficient execution of the project in the field.

Utilities and services: Field office, snow removal, sanitary facilities, telephones, signage, and fencing. These services are necessary to support the field personnel and provide a safe and secure project environment.

Equipment: Lifting equipment, scaffolding and ladders, elevators, trash chute, or temporary doors—the cost of any equipment provided to the job on the whole so as to benefit all field personnel, subcontractors, and management. This equipment should not be priced by individual subcontractors.

Field office costs: The cost of prints or supplies used by field office or drafting personnel, engineers, or other outside consultants needed to support or advise the project management team.

Legal and insurance: The cost of bonds, insurance, and liens required for the project would also be an added general conditions costs.

Cleanup: The cost involved in day-to-day cleanup, rubbish removal, and the final cleanup required when the project is turned over to the owner must also be priced and would be included in general conditions.

These general conditions items would be individually tabulated using a form like the Means project overhead summary illustrated in Chapter 6, Figure 6–17.

Home office overhead is a factor of the volume of the company as compared with the costs associated with running the business. A company reduces its home office overhead by reducing the home office's costs, or by increasing its volume of work. The amount of home office overhead required is dependent on the type of work and the competitiveness of the environment. Highly technical work or regulated work as well as highly competitive or litigious environments would all demand higher home office support.

An example of how home office costs might be calculated can be shown. Assume a construction company with a $2 million yearly volume. What would its home office overhead rate be as a percent of dollar volume?

Office Overhead:

Owner/Engineer/Project Manager	$70,000
(same person)	
Estimator	$50,000
Secretary/Receptionist/Bookkeeper	$30,000
Office/Rent and Utilities	$20,000
Office Equipment	$12,000
Accountant (retainer)	$ 2,000
Legal (retainer)	$ 2,000
Medical, Workmans' Comp.	$22,000
Advertising	$ 2,000
Vehicle	$10,000
Association fees	$ 1,000
Seminars and Travel	$ 2,000
Entertainment	$ 2,500
Bad Debt	$20,000
Total Office Overhead	$245,500

$$\text{Home Office Overhead (\%)} = \frac{\text{Home Office Overhead Costs}}{\text{Company \$ Volume}}$$

$$= \frac{\$245,000}{\$2,000,000}$$

$$= 12.25\%$$

Optimum Project Duration Example

The first step in determining the optimum duration for a project is to prepare a network schedule and an estimate for the project. The schedule and the estimate define the normal cost for the project. The network schedule in Figure 10–6 establishes a normal duration of 37 days for the renovation of an office.

The normal cost would be determined by the estimate, which is broken down by activity in column 4 in Figure 10–7. These costs define the direct costs for the project; combining them with the project's indirect cost establishes the total cost for the

Figure 10–5
Relationship between direct and indirect costs.

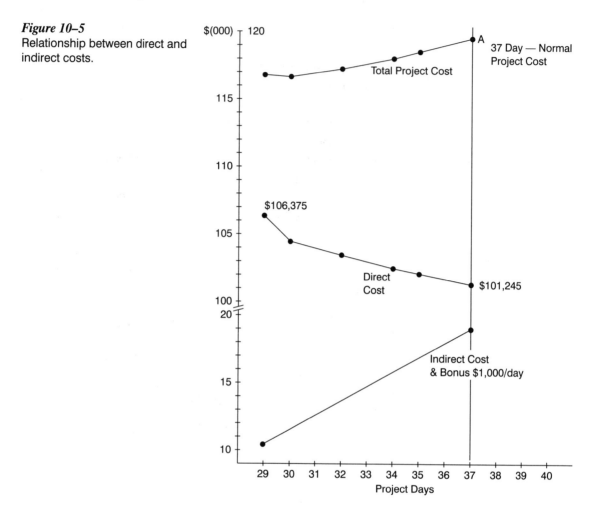

project. This project's total indirect, direct, and total cost for a normal duration of 37 days are marked on Figure 10–5 as point A.

The indirect costs for this project are calculated as follows:

Field supervision	$300/day
Field office and supplies	$50/day
Field equipment	$50/day
Home office support	$100/day
Total	$500/day

Assuming a $500 per day bonus for early completion, the contractor receives a $1,000 per day benefit for every day the project is shortened.

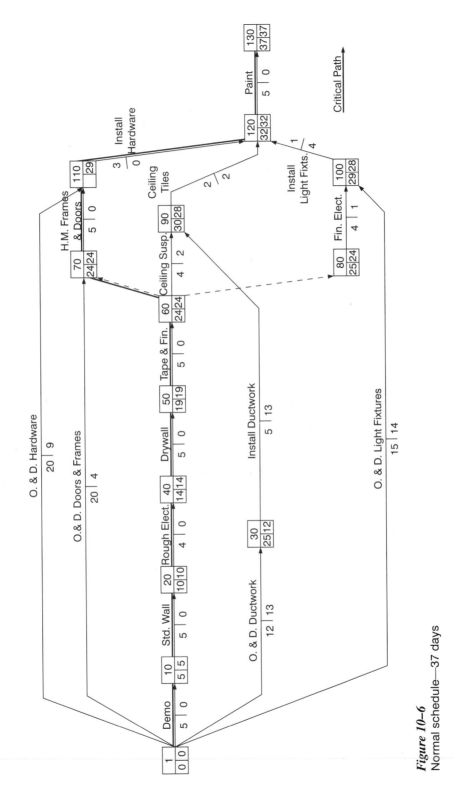

Figure 10–6
Normal schedule—37 days

$i-j$	Description	Normal Duration	Normal Cost	Crash Duration	Crash Cost	Potential Days Saved	Cost per Day
1-10	Demo	5	3,000	3	4,000	2	500
1-30	Order & Deliver Ductwork	12	6,200	—	—	—	—
1-70	Order & Deliver Doors & Frames	20	8,400	—	—	—	—
1-100	Order & Deliver Light Fixtures	15	8,550	—	—	—	—
1-110	Order & Deliver Hardware	20	5,900	—	—	—	—
10-20	Install Stud Wall	5	9,450	3	11,800	2	1,175
20-40	Rough Electric	4	7,500	3	8,600	1	1,100
30-90	Install Ductwork	5	6,000	3	7,600	2	800
40-50	Install Drywall	5	10,200	3	11,900	2	850
50-60	Tape and Finish Drywall	5	6,300	3	7,700	2	700
60-90	Install Ceiling Suspension	4	8,250	3	9,000	1	750
70-110	Install Frames and Doors	5	2,550	3	3,870	2	660
80-100	Finish Electric	4	3,360	3	4,110	1	750
90-120	Install Ceiling Tiles	2	6,450	1	7,500	1	1,050
100-120	Install Light Fixtures	3	3,375	2	4,275	1	900
110-120	Install Hardware	3	1,560	2	2,040	1	480
120-130	Paint	5	4,200	3	5,040	2	420
	Total		101,245				

Figure 10–7
Cost breakdown.

Next, the critical path activities that can be shortened at a daily rate that is less than the daily indirect rate for the project must be identified. In the case of a network with only one critical path, only one activity must be shortened. If a network had more than one critical path, more than one activity would have to be crashed. As activities are shortened, more critical paths will be created.

In this example the least expensive critical path activity to shorten is the Painting activity. This activity can be shortened 2 days at a direct cost of $840. The next least expensive activity is the Hardware Installation activity, which can be shortened one day at a cost of $480. Referencing the new schedule in Figure 10–8, note that the project's duration is now 34 days and two critical paths have been created.

The next least expensive activity to accelerate is the Installation of the Frames and Doors at a cost of $660 per day, but crashing it still would not shorten the job since two critical paths exist at this point in the schedule.

The next best option is to accelerate the Demolition activity 2 days at a cost of $500 per day and after that the Tape and Finish activity, also 2 days at a cost of $700 per day. By crashing these two activities the Order and Delivery of the Doors and Frames has now become critical, and since this activity cannot be shortened further, no more time can be saved in this part of the network. The project's duration is now 30 days, as seen in Figure 10–9.

The next best option for shortening the project is to crash both the Installation of the Frames and Doors as well as the Finish Electric work one day at a combined cost of $1,410. Notice, however that this is more expensive than the indirect cost for the project and that the project's total cost now increases, as noted in Figure 10–10.

Optimization Conclusion

Although there is a clear benefit to optimizing a project's duration on the basis of cost, this is not a routine step in project planning. The integration of scheduling and estimating information cannot be easily linked since the activity units are often not the same. It is also unusual to calculate crash costs for each activity and then to formally analyze and compare crash costs with indirect costs. This process takes a considerable amount of time and is difficult to automate. Another real concern is that as a project is crashed multiple critical paths are created, and as more critical paths appear, the greater the risk of a delay in the completion of the project.

The above reasons notwithstanding, the process of determining the optimum duration for a project is an important step in the proper planning of a project. Considerable time and money can be saved by properly analyzing the costs of a project and then running the project in the most cost effective way. Figure 10–11 displays an optimization study done prior to the construction of an ambulatory care facility. As expert system technology becomes better developed and cost and schedule information becomes more fully integrated, this kind of study will become more routine. (See sidebar by Kenneth H. Stowe on Advanced Project Management.)

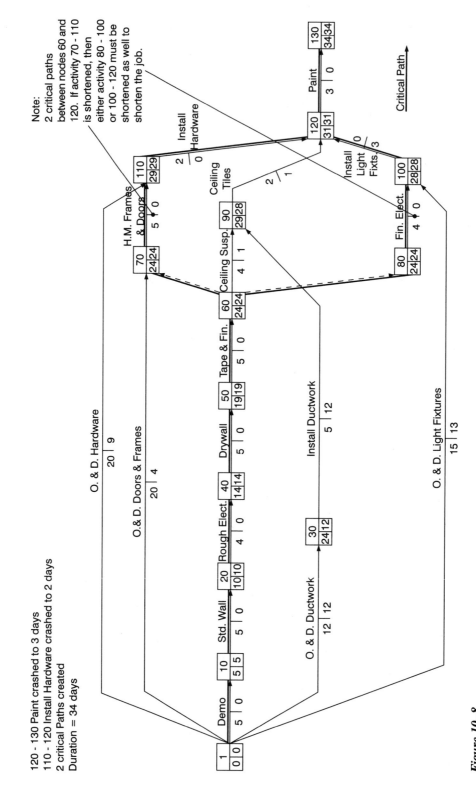

120 - 130 Paint crashed to 3 days
110 - 120 Install Hardware crashed to 2 days
2 critical Paths created
Duration = 34 days

Note:
2 critical paths
between nodes 60 and
120. If activity 70 - 110
is shortened, then
either activity 80 - 100
or 100 - 120 must be
shortened as well to
shorten the job.

Critical Path

Figure 10–8
Expedited schedule 1.

264

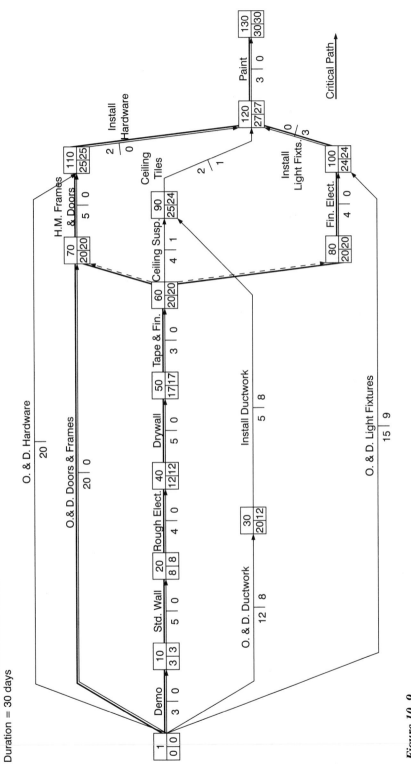

1 - 10 Demo crashed to 3 days
50 - 60 Tape and Finish crashed to 3 days
O. & D. Doors and Frames now critical
Duration = 30 days

Figure 10–9
Expedited schedule 2.

	Direct Cost	Indirect Cost (less Bonus)	Total Cost	Duration
Normal	101,245	18,500	119,745	37
Expedite Paint (2 Days)	102,085	16,500	118,585	35
Expedite Hardware Installation (1 Day)	102,565	15,500	118,065	34
Expedite Demo (2 Days)	103,565	13,500	117,065	32
Expedite Tape and Finish (2 Days)	104,965	11,500	116,465	30
Expedite Frames and Door Installation and Finish Electric (1 Day Each)	106,375	10,500	116,875	29

Figure 10–10
Cost breakdown of expedited schedules.

Optimizing Project Duration

Ambulatory Care - with a Per Month Incentive of $100,000.00

		Project Duration (in Months)				
		21	22	23	24	25	26
Fixed Costs Design		$600,000	$600,000	$600,000	$600,000	$600,000	$600,000
Time-Variable Costs Design		$210,000	$220,000	$230,000	$240,000	$250,000	$260,000
Premium-Design		$35,000	$20,000	$10,000	$0	$0	$0
Direct Costs - Construction		$12,000,000	$12,000,000	$12,000,000	$12,000,000	$12,000,000	$12,000,000
Premium-OT		$220,000	$80,000	$30,000	$0	$0	$0
Premium-Planning		$57,686	$43,256	$32,443	$0	$0	$0
Premium -CADD + Expedite Procurement		$160,000	$40,000	$5,000	$0	$0	$0
Indirect (Time-Variable)		$450,000	$500,000	$550,000	$600,000	$650,000	$700,000
Indirect (One-Time)		$150,000	$150,000	$150,000	$150,000	$150,000	$150,000
Indirect (Premium Staffing)		$60,000	$20,000	$5,000	$0	$0	$0
Owner's Staffing/Supervision		$525,000	$550,000	$575,000	$600,000	$625,000	$650,000
	Total Without Advantage Considerations	$14,467,686	$14,223,256	$14,187,443	$14,190,000	$14,275,000	$14,360,000
Time-to-Market Advantage		($300,000)	($200,000)	($100,000)	$0	$300,000	$600,000
Productivity Advantage		$0	$0	$0	$0	$100,000	$200,000
	Total With Advantage Considerations	$14,167,686	$14,023,256	$14,087,443	$14,190,000	$14,675,000	$15,160,000

Savings with Advantage Considerations $166,744

Figure 10–11a

Compliments of Kenneth H. Stowe, P.E., of the George B. H. Macomber Co., Boston, MA

Optimizing Project Duration

Ambulatory Care - with a Per Month Incentive of $200,000.00

	Project Duration (in Months)					
	21	22	23	24	25	26
Fixed Costs Design	$600,000	$600,000	$600,000	$600,000	$600,000	$600,000
Time-Variable Costs Design	$210,000	$220,000	$230,000	$240,000	$250,000	$260,000
Premium-Design	$35,000	$20,000	$10,000	$0	$0	$0
Direct Costs - Construction	$12,000,000	$12,000,000	$12,000,000	$12,000,000	$12,000,000	$12,000,000
Premium-OT	$220,000	$80,000	$30,000	$0	$0	$0
Premium-Planning	$57,686	$43,256	$32,443	$0	$0	$0
Premium -CADD + Expedite Procurement	$160,000	$40,000	$5,000	$0	$0	$0
Indirect (Time-Variable)	$450,000	$500,000	$550,000	$600,000	$650,000	$700,000
Indirect (One-Time)	$150,000	$150,000	$150,000	$150,000	$150,000	$150,000
Indirect (Premium Staffing)	$60,000	$20,000	$5,000	$0	$0	$0
Owner's Staffing/Supervision	$525,000	$550,000	$575,000	$600,000	$625,000	$650,000
Total Without Advantage Considerations	$14,467,686	$14,223,256	$14,187,443	$14,190,000	$14,275,000	$14,360,000
Time-to-Market Advantage	($600,000)	($400,000)	($200,000)	$0	$300,000	$600,000
Productivity Advantage	($120,000)	($100,000)	($50,000)	$0	$100,000	$200,000
Total With Advantage Considerations	$13,747,686	$13,723,256	$13,937,443	$14,190,000	$14,675,000	$15,160,000

Savings with Advantage Considerations — $466,744

Figure 10–11b

268

Sidebar ▬▬▬▬▬▬▬▬▬▬▬▬▬▬▬▬▬▬▬▬

Advanced Project Management:

Art and Science of Interdependent Disciplines

Advanced project teams can deliver projects faster, for less cost, better suited to the client's needs, and with a richer electronic model for future maintenance. By advanced project management we mean comprehensive critical path schedules with resource data generated by an estimate with a common structure. We also mean extended use of the computer-aided design data during the procurement, coordination, and construction processes, producing a rich electronic source of information for the facilities' operation and maintenance.

New time-to-market pressures in many industries are driving project teams to seek economical ways to deliver quality projects faster. Especially urgent are projects in the medical, pharmaceutical, computer, communications, and entertainment industries.

The advances in project management in the 1990s have been less quantitative and less microscopic than the earned value techniques refined by the megaprojects of the 1970s and 1980s. Those techniques gave us ways to predict speed and measure progress. New advances focus on technology and communication and synergy between disciplines. New computer models of projects are more integrated and more nimble. Creative teams are open and democratic, and focused on group goals sought with concurrent engineering rather than a linear process. The teams are connected by the Internet.

Heretofore, the team would be charged with identifying the most economical solution: a mix of the best materials and methods and a balance of acceleration costs and risks, taking into account the time-variable costs, such as rented cranes, scaffolding, and project personnel (see Fig. A).

The modern project team gives a greater respect for time-to-market advantage. This changes the dynamic of the formula and puts more pressure on teamwork (see Fig. B).

It means "the project schedule" becomes many variations of a baseline schedule model, each reflecting ascending levels of acceleration investment. It means the cost estimator must produce many estimates, one for each of several schedule scenarios. It means the design process must be more tightly connected between disciplines, using e-mail to flash design alternatives and changes to the team. It requires a new look at the benefits of three-dimensional CAD models and their potential to speed decisions and reduce crippling delays and rework (see Fig. C).

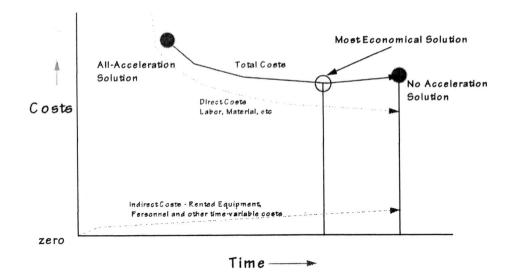

Figure A
Cost-time curves showing most economical solution, ignoring the strategic advantage of early competition.

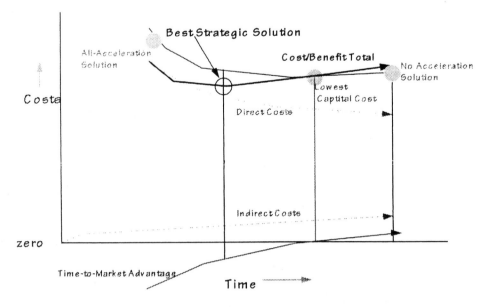

Figure B
Benefit-time curves showing that best strategic decision is faster, the additional expense of which is justified by time-to-market advantages over competitors.

Figure C
5 steps to technologically
excellent.

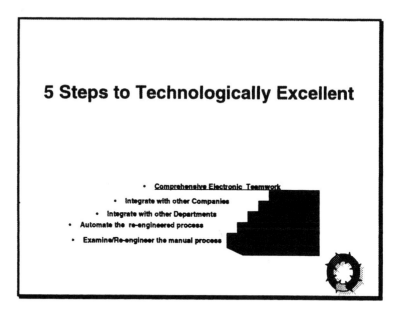

Five Steps to Excellence

Advancing technically in any discipline can be described as a succession of five steps:

1. The first step is examining and reengineering the existing procedures. In the case of estimating, this step likely includes a look at data structure, levels of detail, audience analysis, sources of input, sources of feedback, etc.

2. Rebuilding the process to streamline and enable new technologies is next.

3. The company then addresses more of a political issue than a technical one: integration across departments. Typical of this step is a process such as the designer who asks departments to follow computer-aided design standards and share data.

4. The advanced company then creates links with partners, clients, subcontractors, vendors, permit agencies, etc. At the outset, this may be simply electronic mail of meeting minutes. It can lead to simultaneous access to a secure, detailed project model that is the sum of many disciplines.

5. Finally, the company tackles comprehensive electronic teamwork or enterprise workflow. Now the project goals are optimally served by the technology. Communication is enhanced. Stored wisdom is

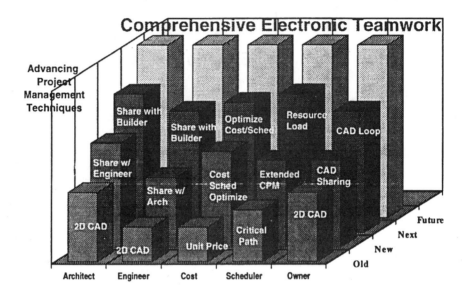

Advancing Disciplines in the Facilities Engineering

Figure D
Improved performance of one discipline can only be achieved with advanced neighboring disciplines.

maximized and redundancy minimized. The advanced team that achieves this level delivers the project at precisely the time justified by the time-to-market urgency. They specify all the scope that the client can afford with a maximum return on investment. All activities add value in pursuit of project goals.

Interdependence

In project management, the more advanced you become, the more your performance is restrained by the "neighboring disciplines" that you may or may not have had a hand in selecting. Consider an advanced planner who employs CPM techniques and electronic mail, who naturally wants to build an integrated project plan for all the disciplines to share. The planner's effort to build cost-and resource-loaded schedules can be hindered by manual, ill-detailed, or unstructured estimating.

Consider an owner who wants the project team to share data over the Internet, expecting savings in time and money. Sharing data can be hindered if the adjacent disciplines don't have compatible, regularly checked e-mail.

Sacrifices

Integrated schedules and performance measurement of design activities can be resisted by territorial individuals. An advanced project manager must get all the disciplines to communicate electronically, share vital data, and coordinate via a common work breakdown structure. This often means sacrifice within one discipline for the sake of the project. Partnering can help ease the strain of this type of sacrifice and keep the team focused on the project goals.

As one of the neighbors tries to advance his practice, he realizes he can be restrained by the other members (see Fig. D).

As such, you may find yourself as you advance in skill being as highly selective in the professionals you choose as partners as you are in choosing your employees and technology.

Kenneth H. Stowe, P.E.
The George B. H. Macomber Co.,
Boston, MA

Conclusion

Project control can best be illustrated as a continuing cycle through which project managers identify a goal, measure results, analyze, make adjustments, and report results. This is an action based process with a feedback loop that can cycle as often as necessary, depending on the nature of the project. The estimate and schedule establish the cost and timing goals for the project. As the project proceeds, the actual results will be compared to the target dates and costs established by the estimate and schedule. Significant deviations from the plan should be analyzed so that corrections can be made either in the ongoing project or in the company's data base so that future estimates and schedules will not repeat mistakes. Project control should be viewed as a learning process by which project team members exchange information, make adjustments, and record results.

In establishing the initial project plan, the project team should look to integrate the estimate and schedule to arrive at the most optimum schedule and budget for the project. Estimates and schedules are usually prepared independently, but as the final preparations are made for the project, every effort should be made to integrate the two. As projects are shortened, indirect costs are saved while direct costs go up. The optimum duration is the duration at which the project can be constructed for the least cost. This is found by the analysis of the project's critical path(s) and crashing—that is, shortening—critical activities as long as the direct cost of shortening the project is less than the indirect cost of the project.

Chapter Review Questions

1. Control requires a system of information tailored to the specific needs of the company.

    __ T   __ F

2. Every key position in an organization needs to have specific goals to meet within a specific time period.

    __ T   __ F

3. Because of the implementation of new technologies, companies have less of a need for developing accurate historical data bases.

    __ T   __ F

4. A crash duration is the shortest time in which an activity can be accomplished by using a larger crew, overtime, extra shifts, or any combination of these devices.

    __ T   __ F

5. Crashing a project forces a company to lose both time and money.

    __ T   __ F

6. Control techniques can be better understood and applied if control is equated with:
    a. Planning
    b. Action
    c. Budgeting
    d. None of the above

7. A project control system must encompass:
    a. Planning
    b. Monitoring
    c. Analysis
    d. Historic data collection
    e. All of the above

8. How can a project manager compress a schedule?
    a. Revise the schedule's logic
    b. Use overtime
    c. Use substitute materials or equipment
    d. Increase the crew size
    e. All of the above

9. Combining the direct cost curve and the indirect cost curve creates a third curve called the:
    a. Total cost curve
    b. Forecast curve
    c. Functionalistic curve
    d. Crash cost curve
    e. None of the above

10. The benefits of an effective project control system are:
    a. It identifies problem areas and trends.
    b. It is a communication tool.
    c. It allows project managers to better manage.
    d. All of the above.

Exercises

1. What are the advantages and disadvantages to the establishment of a detailed system of project controls? Discuss the opposite scenario of an insufficient control system.

2. The cost and schedule data for a building project are detailed below. Assume an indirect cost of $900 per day. What is the optimum (least cost) duration for the project?

ACTIVITY	PRECEDED BY	COST		DURATION	
		Crash	Normal	Crash	Normal
A	-	3,100	2,500	5	6
B	-	8,200	7,000	7	10
C	B	7,000	6,500	5	6
D	-	11,800	10,000	10	12
E	A	5,700	4,200	9	12
F	C, D	8,500	8,200	5	6
G	E, F	6,000	5,200	4	5
H	C, D	8,700	7,500	5	8
I	G, H	6,000	5,600	3	4

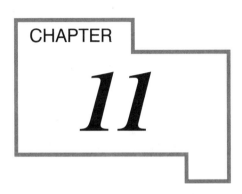
Cost, Schedule, and Resource Control

From studying this chapter, you will learn:

1. The relationship between resources and network schedules
2. How estimates can be integrated with schedules to forecast project value
3. How to perform a cash flow analysis for a project

Introduction

Like a well run company, the well run project is an efficient consumer of resources. What are resources? How are resources utilized in the construction of a project? What are the tools that project managers use to manage these resources so as to ensure that they are used in a cost effective manner? **Leveling**, that is, the process of balancing and smoothly utilizing resources, must be understood in this context, as well as the management of cash flow. When an estimate and schedule are prepared, an unlimited amount of resources are assumed. As this is not usually possible, project managers must examine the collective resource needs of a project and predict and smooth out their consumption. This must be accomplished in order to efficiently run a project and guarantee a profit.

What Are Resources?

To successfully execute a project a tremendous number of resources must be brought together in a coordinated fashion. Labor, equipment, money, and available space are some of the resources that must be procured and managed to complete a project. The estimate quantifies and prices the resources needed, while the schedule creates a roadmap and predicts how long the resources will be needed. Both the estimate and the schedule are created assuming a normal crew size and normal equipment needs.

Labor

In estimating labor requirements, each task is visualized and the trade and crews assigned are based on normal past practice. Each activity is considered independently, and normal working conditions are assumed. Design professionals such as architects or engineers, and tradespeople such as carpenters, electricians, or masons would all be considered labor resources. If unusual local conditions exist—for example, workers are in short supply—then these issues may need to be formally studied.

Equipment

Normal equipment availability is also assumed in the estimating and scheduling of the project. Suppliers of critical equipment such as cranes, trucks, or pavers may need to be contacted to assure their availability. Equipment is an expensive resource for a contractor, so it is important that equipment be efficiently utilized on the project.

Space

Particularly in the case of an urban project, laydown areas and equipment operations can be severely limited by available work area (see Fig. 11–1) . As noted, the estimate and schedule may have assumed normal amounts of room in which to operate. If space is tight, contractor efficiency can be undermined because of the need to constantly coordinate with others on the site. Combined resource needs and availability can be verified and managed through a resource plan.

Cash

Cash is also a resource that must be managed to successfully accomplish a project. Contractors, in particular, need to balance the cash received from the owner for work accomplished with the cash paid out to subcontractors, labor, and material and equipment suppliers.

Why Manage Resources?

The efficient use of project resources is the goal of every project manager. Idle equipment or people are wasteful and will ultimately lead to lost profits. Furthermore, it is expensive to hire and then lay people off, so project managers need to organize the project to provide steady employment for each of the trades required on the project. Equipment mobilization is also expensive, so cranes and pile-driving equipment should be set up once, steadily used, and then moved off the site.

A complete resource study will identify when and how many of each trade will be on the job site each day of the project. This allows management to ensure that work scheduled concurrently is compatible and that adequate supervision is provided. As an example, a sandblasting operation scheduled to occur above some structural steel work would need to be rescheduled, since the sand and removed

Figure 11–1
Available work space on a job site is a resource that must be efficiently managed.
Courtesy of New England Deaconess Hospital and Walsh Brothers, Inc.
Photo by Don Farrell

280

paint would fall upon the steelworkers, causing an unsafe work environment. Also, congested work areas create limits as to how many people or pieces of equipment can be handled at once.

During the estimating stage, the estimator must not only use labor unit prices but also consider how the work will be scheduled in the field. Resource studies can be used to adjust the estimated labor to reflect the true cost of labor.

Consider the following example:

160 | Raceways

		160 200	Conduits	CREW	DAILY OUTPUT	MAN-HOURS	UNIT	MAT.	LABOR	EQUIP.	TOTAL	TOTAL INCL O&P	
205	1030	1-1/2" diameter		1 Elec	65	.123	L.F.	2.80	3.38		6.18	8.20	205
	1050	2" diameter			60	.133		3.85	3.67		7.52	9.75	
	1070	2-1/2" diameter			50	.160		6.20	4.40		10.60	13.40	
	1100	3" diameter			45	.178		8.25	4.89		13.14	16.40	
	1130	3-1/2" diameter			40	.200		10.20	5.50		15.70	19.45	
	1140	4" diameter			35	.229		12.10	6.30		18.40	23	
	1750	Rigid galvanized steel, 1/2" diameter			90	.089		1.25	2.44		3.69	5.05	
	1770	3/4" diameter			80	.100		1.55	2.75		4.30	5.85	
	1800	1" diameter			65	.123		2.10	3.38		5.48	7.40	
	1830	1-1/4" diameter			60	.133		2.65	3.67		6.32	8.40	

From *Means Building Construction Cost Data 1995*. Copyright R. S. Means Co., Inc., Kingston, MA, 617-585-7880, all rights reserved.

The estimated cost of labor for installing 300 feet of conduit, using the unit price of $2.75 (see above table) for labor is

$$\text{labor cost} = 300 \text{ ft} \times \$2.75/\text{ft}$$
$$= \$825$$

Let's look at the scheduled cost. Referencing the above table one electrician can install 80 lf of conduit per day; therefore, one electrician must be assigned for 4 days or 2 electricians for 2 days. Referencing Table 6–12, an electritian earns $27.50/hr would be

$$\text{labor cost} = 2 \text{ days} \times 16 \text{ hr per day} \times 27.50 \text{ per hr}$$
$$= \$880$$

The difference between the two costs is that the $2.75/ft unit price estimates the exact cost of the conduit, assuming that the electritian can move on to other tasks. The scheduled cost estimates a full day's pay for the electricians, assuming some inefficiency.

Even when creative crew assignments are made and activities are combined to maximum efficiency, actual project costs will exceed the estimated costs. Contractors can increase efficiency by splitting workers' time between multiple projects or, in the case of flexible work rules, using workers for a variety of tasks.

Labor and Equipment Studies

To examine labor and equipment requirements for a project, the resource requirements of each activity must be identified. This is done as each individual activity is scheduled. Normal crew and equipment assignments are made with the goal of accomplishing each activity in the most cost efficient manner. These crew assignments can be noted above each activity on the network diagram, or, in the typical case where a computer scheduling system is employed, resource requirements can be attached to each individual activity.

Once resource requirements are assigned to each activity, it is possible to examine the resource needs for the project on any given day. As shown in Figure 11–2, the resource needs for a project can be looked at if each activity is scheduled to start on its early start date, or at the late start date as shown in the second plot. In either case the project manager can see what the peak resource requirements are for each resource studied. Individual resources can be studied as well as cumulative plots which show the total resource requirements for the project.

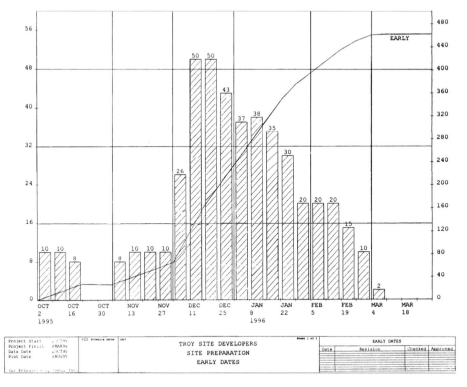

Figure 11–2a
This graphic was created using Primavera Project Planner® (P3®), a product of Primavera Systems, Inc.

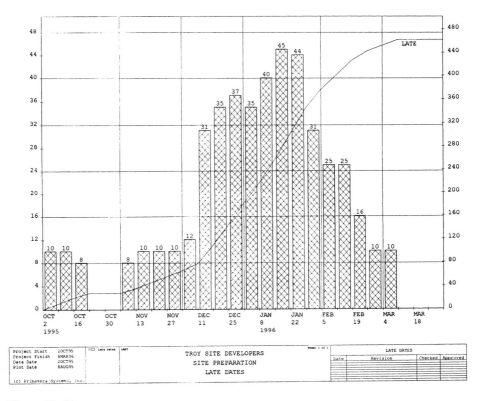

Figure 11–2b

This graphic was created using Primavera Project Planner® (P3®), a product of Primavera Systems, Inc.

As can be seen in the graphs, the resource requirements for the project vary from one day to the next. If the job was managed in this manner, people and equipment would be forced on and off the project and the costs incurred would be excessive. It is expensive to hire and lay off workers as well as to mobilize and demobilize equipment. The ideal situation is for a project to smoothly build up its resources and then to slowly lower its resource requirements as the job closes down. Figure 11–3 illustrates this graphically.

Float is a tool available to project managers to smooth out, or level, their resource requirements. By using float, activities can be manipulated without delaying the ultimate completion of a project. Figure 11–4 illustrates the use of float to smooth out resource requirements.

The process illustrated in that figure would be performed for each resource under consideration. This is a time-consuming calculation and would normally be done with the assistance of a computer. However, even with the computer the process usually is applied only to the most critical resource need. The time involved in entering data into the computer and then analyzing and revising the results can be

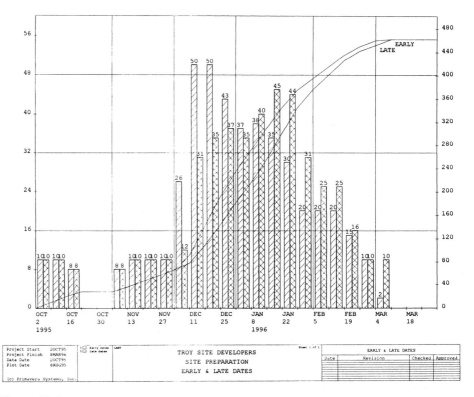

Figure 11–2c
This graphic was created using Primavera Project Planner® (P3®), a product of Primavera Systems, Inc.

Figure 11–3
Project resource effort.

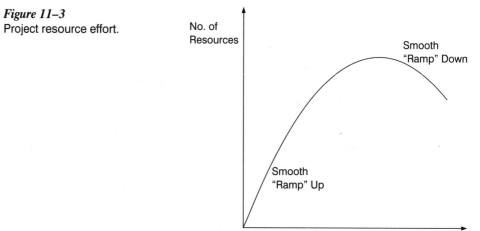

Figure 11–4
Resource leveling.

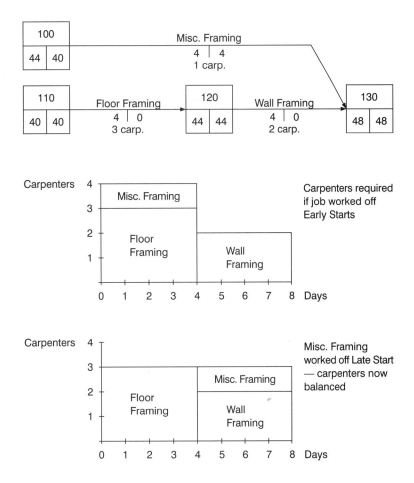

Figure 11–4
Resource leveling.

considerable. A design office might study their design staff's workload, or a construction company might look at tower crane activities or study the available laydown areas.

When using a computer, the leveling process is accomplished by first identifying the resource to be studied. A resource limit such as one tower crane or 10 carpenters will be identified. When the leveling function is executed, the computer will try to move activities around using the available float to accomplish the project without exceeding the resource limits and also not extending the project's end date. If this is not possible, the project will either have to be extended, or the resource levels will have to be increased. When the leveling calculation is completed, a report will be printed identifying the optimum start and finish dates for each activity (see Fig. 11–5). These start and finish dates provide a smooth usage of the resource while still getting the project done on time.

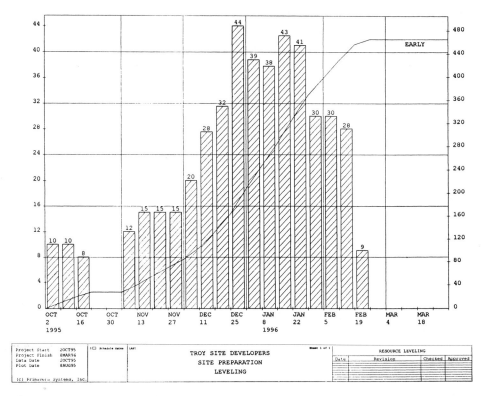

Figure 11–5a
This graphic was created using Primavera Project Planner® (P3®), a product of Primavera Systems, Inc.

Cash Flow Analysis

Cash, like people and equipment, is also a resource that must be managed on a project. It is the rare owner who begins a project with an amount of cash equivalent to the project budget sitting in the bank waiting to be spent on the project. The owner needs to know, with accuracy, how much cash must be available each month of the project to pay the contractor's invoices. Like the owner, the contractor also needs to be able to predict its cash needs for a project. Contractors receive income from the owner in the form of paid invoices. That cash is then paid out to in-house labor, subcontractors, and material and equipment suppliers. To stay in business, the contractor must diligently manage these **"cash flows."**

The management of cash flow is made difficult by the fact that payments are made in different increments depending on the type of activity. Subcontractors generally invoice the general contractors at the end of each month for the work completed. The general contractor typically pays the subcontractor after the general

TROY SITE DEVELOPERS

PRIMAVERA PROJECT PLANNER

REPORT DATE 08AUG95 RUN NO. 30
18:05

RESOURCE LOADING REPORT

RESOURCE LOADING REPORT

TOTAL USAGE FOR WEEK

LABORER - ( )

ACT ID	DESC	TOTAL	2OCT 1995	9OCT 1995	16OCT 1995	23OCT 1995	30OCT 1995	6NOV 1995	13NOV 1995	20NOV 1995	27NOV 1995	4DEC 1995	11DEC 1995	18DEC 1995	25DEC 1995	1JAN 1996	8JAN 1996	15JAN 1996	22JAN 1996	29JAN 1996	5FEB 1996	12FEB 1996	19FEB 1996	26FEB 1996	4MAR 1996
005	CLEAR SITE	28	10	10	8																				
015	ROUGH GRADE	44						8	10	10	10	6													
020	DRILL WELL	25										2	5	5	5	5	3								
025	INSTALL WELL PUMP	24															4	10	10						
030	UNDERGROUND WATER PI	19																5	5	5	3				
040	WATER TANK FOUND.	56										8	20	20	8										
045	ERECT WATER TANK	40													6	10	10	10	4						
050	TANK PIPING	20																	3	4	5	5	2		
055	EXCAVATE FOR SEWER	40										4	10	10	10	6									
060	INSTALL SEWER	45											10	6	6	15	6								
065	EXCAVATE FOR MANHOLE	20										4	10	4	10										
070	INSTALL MANHOLES	30												5	4	6		4							
075	INSTALL ELEC. DUCT B	26										2	5						10	2					
080	OVERHEAD POLE LINE	16																		10	10				
085	PULL IN POWER FEEDER	30																			8	10	10	10	2
TOTAL	LABORER	462	10	10	8			8	10	10	10	26	50	50	50	43	37	38	35	30	20	20	20	15	10	2
	REPORT TOTAL	462	10	10	8			8	10	10	10	26	50	50	50	43	37	38	35	30	20	20	20	15	10	2

Figure 11-5b
Optimum start and finish dates, by activity.

287

PROJECT SCHEDULE

PROJECT: Repair Garage

DATE: 01-04-93

BY: PLJ

NO.	DESCRIPTION	$	COMMENTS ($1,000's)
1 - 12	Clear and grub	$2,979	210
12 - 14	Exc. ftg and utils.	879	195
14 - 16	Form footing	2,629	180
15 - 24	Underground piping	6,184	165
16 - 18	Reinf. and conc. ftgs.	1,973	150
18 - 20	Foundation block	5,527	135
20 - 24	Perimeter insulation	760	120
25 - 28	Gravel fill	714	105
24 - 26	Backfill	1,437	90
26 - 29	Complete backfill	826	75
28 - 30	Masonry bearing walls	54,496	60
29 - 30	Rough plumbing	3,197	45
30 - 36	Erect prestressed conc. slabs	2,575	30
32 - 38	Fine grade, forms and mesh	2,497	15
36 - 38	Erect joists and deck	11,785	0
38 - 42	Concrete floor slab	6,790	
40 - 44	Roofing, sheet metal and skylights	15,489	
42 - 43	Block partitions	1,643	
43 - 54	Toilet fixtures	5,141	
44 - 62	Roof heating units	23,975	
46 - 62	Overhead doors	8,372	
43 - 54	Wire mesh partitions	1,947	
50 - 54	Erect steel stair	1,686	
52 - 54	Rough and finish carpentry	3,949	
54 - 62	Paint	7,020	
56 - 62	Toilet partitions	571	
58 - 62	Resilient flooring	1,507	
60 - 62	Power and lighting	25,791	
62 - 64	Cleanup	0	
		$202,339	

	MARCH	APRIL	MAY
$ Completed per month	$ 20,171	$ 78,287	$ 103,881
-10% Retainage	$ 2,017	$ 7,829	$ 10,388
Monthly payment	$ 18,154	$ 70,458	$ 93,493

Figure 11-6

Cash flow projections, by activity.

From *Means Scheduling and Project Management Workbook.* Copyright R. S. Means Co., Inc., Kingston, MA, 617-585-7880, all rights reserved.

contractor's invoice has been paid by the owner. A general contractor pays its labor at the end of each week. Materials are generally provided to the contractor on credit, with the contractor paying the supplier in full at the end of the month. To properly project cash flow, each activity must be tracked in the manner in which the payment will be made.

The technique that is used to project cash flow is very similar to the manner in which resources are managed. The cost of each activity must first be identified from the estimate and assigned to the activity. If this is being done manually, the network can be plotted in a time-scaled bar chart form and with the activity costs noted on each activity as shown in Figure 11–6.

Each activity listed in that figure shows the duration, cost, and cost slope. The cost slope of an activity is calculated by dividing the cost of an activity by its duration. It is the cost of the activity per duration unit. Once the cost of each activity is assigned, it is possible to add up the cost of the work per payment period, as shown at the bottom of Figure 11–6. This figure is commonly called a cost-loaded schedule; it is the basis for projecting both income and costs for project managers.

Income Projection

A **schedule of values curve**, the name commonly used to describe the cost-loaded schedule, projects the value of work that is scheduled to be invoiced at the end of each payment period. The income received by the contractor is the amount of the invoice less retainage. **Retainage** is money held back by the owner until the contractor satisfactorily completes the contract. Retainage of 5–10 percent of the amount invoiced is normal. As can be seen in Figure 11–7, the values on the income plot equal the values in the schedule of values curve, less retainage. The income plot is a step curve reflecting the fact that the contractor will submit an invoice at the end of each payment period equal to the schedule of values curve and will receive a payment 3 weeks later equal to that value, less retainage. No income is received until the next invoice is paid one payment period later. All retainage is paid to the contractor at the end of the project. The final point on the schedule of values curve must equal the income curve, which equals the contract amount.

The projected payment period amounts are commonly negotiated between the contractor and the owner. It is not unusual for the contractor to shift cash demands to the front of the job, overvaluing early activities while undervaluing later items. This is commonly called **front end loading** the job. Within reason, this is an accepted practice since in the early part of a project contractors often are faced with hidden costs not easily attached to specific work activities. This also helps offset owner retainage, which can severely impact a contractor's cash flow.

When assigning a cost to an individual activity it is important that the amount attached reflects the contractor's complete costs, including both direct and indirect costs as well as profit. Subcontractor costs must be marked up, and material and labor costs must also include appropriate markups. The costs assigned to each activity are what will eventually be billed to the owner; therefore all costs and profit must be identified.

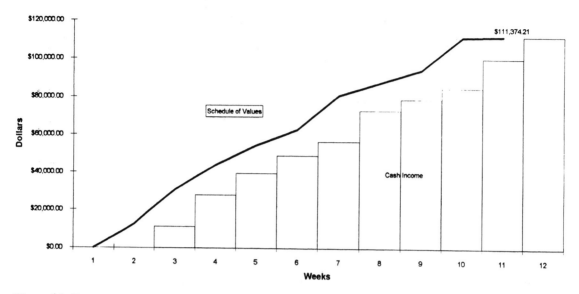

Figure 11–7

Payment Projection

A **payment curve**, also called a **cash requirements curve**, should also be prepared by the contractor. This curve projects the cash payables expected by the contractor for the project. This curve is prepared in similar fashion to the income curve, except that now the contractor is forecasting the cash that is leaving the company to pay for labor, materials, and subcontracts. Another difference between this curve and the income curve is that the costs assigned to each activity should be the direct costs, not including general overhead or profit. At the end of the project the final point on the cash requirements curve will indicate the total amount of money that the contractor spent for labor, materials, equipment, and subcontracts, as the income curve reflects the total amount of money paid to the contractor by the owner. The difference between the two curves is the money necessary to pay for general overhead and provide a profit.

The basis for the cash requirements curve is the cost curve, which is a summary of all payment categories such as labor, materials, and subcontractors on a periodic basis, usually monthly. The direct cost of each item is taken from the estimate, and the placement of each activity is taken from the schedule. In plotting the cost curve the period costs are added up and plotted in an **S-curve** as shown in Figure 11–8. The production curve, also shown in that figure, is plotted similarly. The production curve values include direct cost, overhead, and profit.

The cost curve thus identifies the direct cost of the project at any point in time. This information can be used to project payroll requirements, as well as material

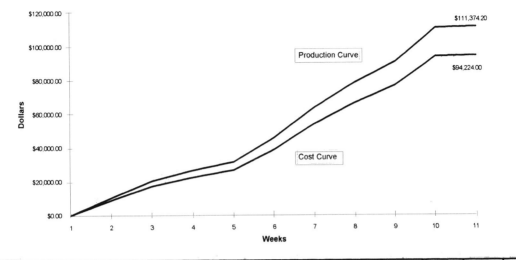

Cost & Production Curves

	5	10	15	20	25	30	35	40	45	46
Duration	5	10	15	20	25	30	35	40	45	46
Calender Days	4/3 4/10	4/10 4/17	4/17 4/24	4/24 5/1	5/1 5/8	5/8 5/15	5/15 5/22	5/22 5/29	5/29 6/5	6/5 6/6
1 Labor	500.00				5,800.00	3,335.00	1,339.00	647.00	1,345.00	600.00
2 Material							3,500.00	1,284.00	1,600.00	
3 Subcontractors	8,200.00	8,400.00	4,433.00	5,067.00	1,200.00	11,938.00	738.00	5,283.00	15,836.00	
4 Suppliers		4,707.00	4,707.00	3,765.00						
5 Period Cost Totals	8,700.00	13,107.00	9,140.00	8,832.00	7,000.00	15,273.00	5,577.00	7,214.00	18,781.00	600.00
6 Cumulative Totals	8,700.00	21,807.00	30,947.00	39,779.00	46,779.00	62,052.00	67,629.00	74,843.00	93,624.00	94,224.00
7 Production value period totals	10,283.53	15,492.67	10,803.62	10,439.56	8,274.11	18,052.92	6,592.10	8,527.06	22,199.43	709.21
8 Cumulative production values	10,283.53	25,776.21	36,579.82	47,019.38	55,293.49	73,346.41	79,938.51	88,465.56	110,664.99	111,374.20

Figure 11–8

expenses for the project. The reason that the costs have been categorized as labor, material, or subcontractor is that each payment category is paid differently. In-house labor is paid weekly, materials are paid for at the end of the month, and subcontractors are paid after the owner pays the general contractor. Therefore, to properly model the cash payables for the contractor, these payment scenarios must be accounted for. Note in Figure 11–9 how the cost curve has been modified to account for the different cost categories. The labor category is drawn as a sloped line to denote that labor is paid weekly, or throughout the payment period. The materials are shown being paid in full at the end of the month. Subcontractors are paid last, in the fourth week of the following period. In some cases the general will also hold back 10 percent retainage from the subcontractor, which could also be modeled if required.

The cash requirements curve shown in Figure 11–9 is an attempt by the contractor to model as precisely as possible its cash needs for a project. The payment categories used in the example are the most common; other payment categories can cer-

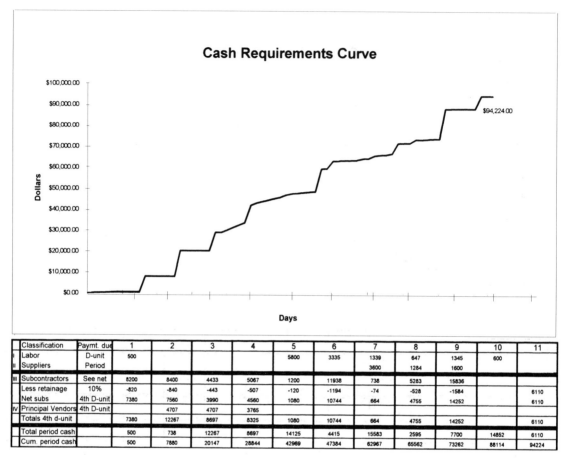

Figure 11–9

Classification	Paymt. due	1	2	3	4	5	6	7	8	9	10	11
I Labor	D-unit	500				5800	3335	1339	647	1345	600	
II Suppliers	Period							3600	1284	1600		
III Subcontractors	See net	8200	8400	4433	5067	1200	11938	738	5283	15836		
Less retainage	10%	-820	-840	-443	-507	-120	-1194	-74	-528	-1584		6110
Net subs	4th D-unit	7380	7560	3990	4560	1080	10744	664	4755	14252		6110
IV Principal Vendors	4th D-unit		4707	4707	3765							
Totals 4th d-unit		7380	12267	8697	8325	1080	10744	664	4755	14252		6110
Total period cash		500	738	12267	8697	14125	4415	15583	2595	7700	14852	6110
Cum. period cash		500	7880	20147	28844	42969	47384	62967	65562	73262	88114	94224

tainly be added as needed. However, the more categories added, the more time-consuming the processing. Other example categories might be material suppliers who provide a discount if payments are made in full within 10 days, or large material vendors who will often provide equipment for installation and not require payment until one week after the contractor receives the owner's payment.

Cash Flow Analysis Conclusion

The computation of both a cash requirements curve and an income curve allows the construction professional to accurately project all cash needs and excesses for the project. This can be best visualized by overlaying these two curves, as illustrated in Figure 11–10.

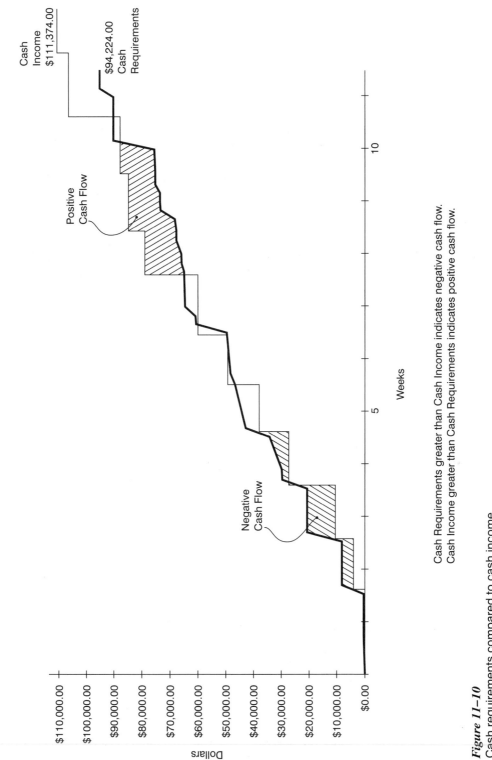

Cash Requirements greater than Cash Income indicates negative cash flow.
Cash Income greater than Cash Requirements indicates positive cash flow.

Figure 11–10
Cash requirements compared to cash income.

293

As can be seen in this plot, this project will need a source of additional financing at its beginning, but as it nears completion the project will begin to produce a positive cash flow. Early negative cash flow is typical for most projects, and must be planned for. Owner retainage as well as the lag in the owner's payment of requisitions is the cause of this negative cash flow.

Contractors can minimize this negative cash by front end loading the job. To front end load a job, early work is billed at a rate higher than its actual cost, while later work is undervalued. Contractors do not pay their subcontractors until after they receive owner payment, as opposed to in-house labor which is paid immediately. Cash flow can be improved by using more subcontractors, as well as by using subcontractors for the early work. Negotiating a lower rate of retainage, freeing up retainage earlier, arranging additional credit from material suppliers, or prearranging a credit line from a bank can all help cover the early cash flow problems.

As float was used to help balance resources, it can also be used to help manage cash flow. Early, high-cost activities that demand contractor investment such as in-house labor or material purchase items can be delayed if they have float. Subcontractor work or deliveries from material suppliers who are extending credit can be scheduled early to help minimize negative cash flow.

Cash requirement and income projections should be done for all company projects since most companies have projects running at different stages of completion. Projects requiring an influx of cash can be helped by other projects which are nearing completion and generating positive cash flow. Most of the newer scheduling software packages allow activities to be cash loaded to produce these kinds of cash studies. Spreadsheets can also be used to summarize overall company cash flow.

Conclusion

This chapter illustrated how schedules can be used to better manage the overall resources of a project. Schedules when originally set up assume an optimum crew and equipment size for each activity. However, when the project as a whole is examined, it may make sense to adjust the start, finish, or duration of an activity or a project to more optimally utilize the company's resources. It is expensive to move people and equipment on and off the site, so project managers should plan their projects to smoothly build up and then taper down their resource needs.

Cash is also a resource that should be managed. By integrating cost information from an estimate with the timing information of a schedule, the cash needs of a company can be closely examined. Network schedules can be cost loaded to accurately forecast both the cash moving in and that moving out of a company. Cash income and requirements curves can be compared to clearly show when and for how long a project will be running cash negative. This allows project mangers to make arrangements in advance to minimize or handle cash shortfalls.

Chapter Review Questions

1. A cost-loaded CPM can be used to project a contractor's cash flow.

    ___ T   ___ F

2. Estimates and schedules, when initially prepared, assume unlimited resources.

    ___ T   ___ F

3. The reason that the scheduled cost of labor may be more than the initial estimated cost is explained by the increase in the unit price of labor.

    ___ T   ___ F

4. Float is a useful tool for project management to use in order to level resources.

    ___ T   ___ F

5. The final point on the cash requirements curve exceeds the final point on the cash income curve by an amount equal to overhead and profit.

    ___ T   ___ F

6.  and cost dimensions can be combined to forecast the rate of spending on a project.
    a. Wage rates
    b. Time
    c. Number of workers
    d. Productivity

7. Which would *not* be considered a scheduling resource?
    a. People
    b. Time
    c. Equipment
    d. Laydown space

8. Given a network schedule presently scheduled for 120 days, a project manager analyzes the carpentry activities and finds that the four available carpenters are not enough to complete the project on time. What options are available to the project manager?
    a. Delay the project
    b. Increase the number of carpenters
    c. Look to revise the network's logic
    d. All of the above

9. What option below would *not* improve the cash flow of a project for a contractor?
    a. Increase front end loading
    b. Subcontract more of the earlier work
    c. Increase owner retainage
    d. Delay in-house labor activities

10. Why might an owner request a schedule of values submittal from the contractor before work is begun?
    a. To verify project quality
    b. To analyze the schedule's float
    c. To verify contractor's payment requests
    d. None of the above

Exercises

1. Shown on pages 297–298 is a precedence diagram for a small building project. Identified below each activity is the number of laborers required for that activity. Given a limit of ten laborers, can the project be completed on time?

2. Using the data in the table, complete Assignments a–d.

Activity	Duration	Total Cost
1-2	4	8,000
1-3	1	2,500
1-6	12	24,000
2-4	10	20,000
3-5	8	16,000
4-6	8	12,000
5-6	8	8,000
5-7	4	10,000
6-8	8	24,000
7-8	7	21,000

a. Draw an *i-j* network and perform the forward and backward pass.
b. Complete the table below:

Activity	D	Cost	Cost/Slope	ES	EF	LS	LF	TF
1-2	4	8,000						
1-3	1	2,500						
1-6	12	24,000						
2-4	10	20,000						
3-5	8	16,000						
4-6	8	12,000						
5-6	8	8,000						
5-7	4	10,000						
6-8	8	24,000						
7-8	7	21,000						

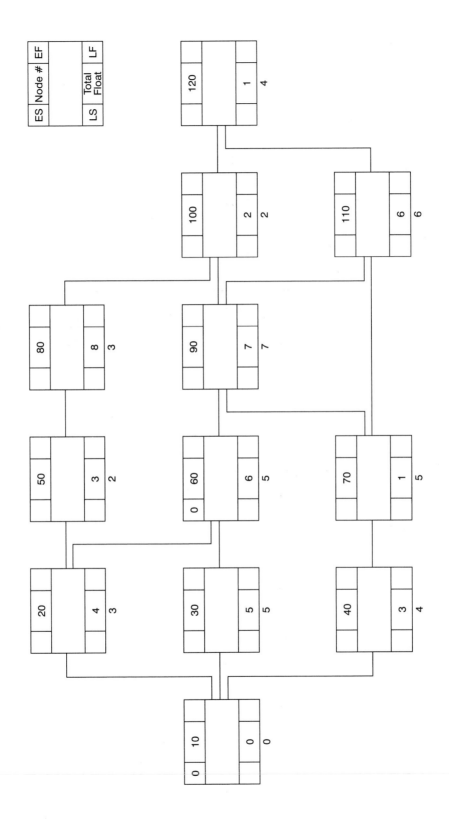

Resource Allocation Table

(Workdays)

Activity	Dur.	R.	R-d	E.S.	T.F.	Priority	1	2	3	4	5	6	7	8	9	10	11	12	13	14	15	16	17	18	19	20	21	22	23	24	25

c. Plot the cost schedule below. Use the early project times.

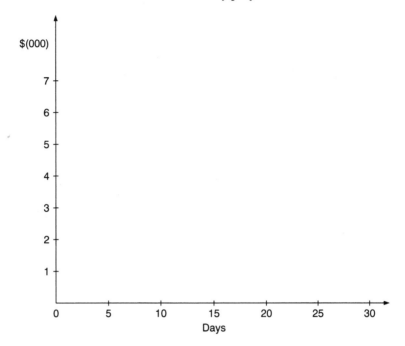

d. Management has stated that the project job cost cannot exceed $6,000 in any one week. Can the project be completed under this constraint without delaying the completion date calculated in Part a above?

CHAPTER

12

Updating the Project:
Control in Practice

CHAPTER OUTLINE

From studying this chapter, you will learn:

1. How to monitor and update a project's budget

2. How to monitor and update a project's schedule

3. How to gather and process actual project data

4. How to properly document a project

Introduction

A project begins with a work plan that includes a budget, a schedule, and an engineered approach designed to complete the project in the most effective manner. The preparation of a work plan involves many project participants, whose goal is to most efficiently utilize the company's resources. The tools used to develop the work plan are the estimate and the schedule, which form the basis of an analysis of the project.

Projects begin with this plan, but the reality of the construction process is that events occur which force the plan to be altered. An owner adds or deletes work, forcing a change in the project's scope; the project experiences bad weather; a labor strike occurs; or the productivity experienced in the field is less than planned. These events all require the project team to establish a system to capture the actual events of the project, analyze them, and make the appropriate adjustments. The project's budget, schedule, and quality must all be controlled to guarantee the success of the project for the owner.

This chapter will look at the use of estimates and schedules as control tools. It will discuss the methods that are used to measure actual progress on a project and how this actual information can be compared to the plan to determine the project's productivity. Examples of job status reports, job analyses, and project forecasting will be included. The intention of the project control process is to examine the project in progress and to intelligently respond to actual events so as to complete the project as planned.

Project Member Viewpoints

Owner

The owner's goals are that the project be delivered with the requested quality, on time, and within budget. As the project proceeds, the owner must receive continued assurance as to the forecasted costs and expected completion date. Each time a requisition is submitted, an owner should know what specific work is being paid for. In a fixed price contract the owner needs to ensure that work is not being overpaid. In a reimbursable contract the owner needs to pay for actual contractor costs, which should be justified. The owner also must project cash flow and the date of completion. As changes occur due to changes in scope or unexpected events, the need for tight control of cost and schedule becomes increasingly important. Keeping the cost within acceptable budget constraints and maintaining control over key milestone completion dates for coordination with outside parties is critical to the owner.

Designer

The designer's involvement with the control process is dependent on the type of delivery method that is selected by the owner. In a traditional method the designer prepares the contract documents completely and then delivers these to the owner before the bidding process begins. The designer's responsibility is to deliver a complete and technically accurate design on time to the owner. Once construction begins, the designer may be hired to oversee the construction. In this capacity the designer would review the quality of the work as well as evaluate progress payments. If changes occur, the designer would assist the owner in negotiating changes in the contract amount as well as the project's completion date.

In a phased or fast-tracked project the design of the work package is by necessity integrated with the bid and construction work. The overall master schedule for the project would link the design periods to the bid and construction work. The project team would include the owner, construction manager, and designer as the three key players. The designer would participate with the owner and construction professional in the selection of contractors, approval of requisitions, and the negotiation of changes.

Construction Manager

A Construction Manager hired onto a project acts as an agent of the owner, and in this capacity, his/her interests are the same as the owner's. Depending on the contract established, the Construction Manager's scope of work would normally include oversight of the contractor and—as a key player in the owner, designer, construction manager model—participating with the owner and architect in contractor selection approvals and negotiations. The management of the master schedule and project budget would also be the Construction Manager's responsibility.

Contractor

In the case of a fixed price contract the contractor maintains tight control over the budget and schedule to guarantee a profit on the project. Documentation of actual progress costs and productivity is important, so as to allow for intelligent midcourse corrections on the project, as well as for the pricing of future projects or the negotiation of claims.

Control Baselines

The creation of a **baseline**, also called a **target** or project forecast, is the first step in the control process. The baseline establishes the goals for the project and allows management to measure how well the project is proceeding and what the end result will be. The estimate serves as the cost baseline for a project while the time baseline is the schedule.

Cost Baseline

The estimate is the basis by which the costs on a project are first established and later refined as a project proceeds. The conceptual estimate, the first estimate performed for a project, establishes the initial cost for the project. As the design of the project proceeds, square foot and assemblies based estimates "tighten up" the budget as the project's design becomes better understood. Once the design of the project is completed, bids are solicited, with the accepted bid price establishing the construction budget for the project. This construction budget would be based upon a detailed estimate prepared by the contractor using a complete design including many subcontractor and vendor quotes.

From the contractor's perspective, the contract price is the initial cost baseline for the project and the target by which the success of the project will be measured. Material quantities, labor unit prices, and assumed productivity rates used in this estimate become the target baselines used to measure the project's success. From the designer's and the owner's perspective the agreed-upon contract price is one more price in a series of prices that have been received since the first conceptual estimate. If the scope of the project has been well managed, and if the estimates have been prepared well, then the bids received should be close to the estimate. The quality of the contract documents, the bidding environment, and the location and type of project are all factors that affect the prices received.

Throughout the design and construction process the application of **Pareto's Law** plays an important part in the control of project costs. Pareto, an Italian economist, taught that 80 percent of the outcome of any project is determined by 20 percent of its included elements. Applied to the construction process, any project control system needs to identify the major cost elements of the project early and develop a system of controls to monitor and manage these elements. Projects are broken down through the CSI format, through the use of a work breakdown structure (WBS—see sidebar Chapter 8), and by the use of bid or work packages. In some cases these may or may not be the same. Project managers focus on the elements that will have the greatest impact on the final project cost, and/or the element with the greatest risk of escalation. Shown in Figure 12–1 is an example of a project cost report which identifies cost and schedule information by project element.

The estimate is clearly the baseline for the control of project costs. As the project moves to the construction stage the estimate becomes extremely detailed, with

```
-----------------------------------------------------------------------------------------------------------
Yankee Construction Company                  PRIMAVERA PROJECT PLANNER              Ruth Building

REPORT DATE  27JUL95  RUN NO.   5            COST CONTROL ACTIVITY REPORT           START DATE  4SEP95  FIN DATE 19OCT95
             17:06
CC-10                                                                              DATA DATE  4SEP95   PAGE NO.   1
-----------------------------------------------------------------------------------------------------------
                     COST      ACCOUNT  UNIT                PCT     ACTUAL    ACTUAL     ESTIMATE TO
ACTIVITY ID RESOURCE ACCOUNT   CATEGORY MEAS    BUDGET      CMP    TO DATE  THIS PERIOD  COMPLETE    FORECAST    VARIANCE
----------- -------- -------   -------- ----  ----------- -----  --------- -----------  ---------- ----------- -----------
  010 Excavate Footings
      RD     6 ES  4SEP95  EF 11SEP95  LS  4SEP95  LF 11SEP95  TF    0

      LABOR                                  10800.00   .0       .00       .00       10800.00    10800.00        .00
                                           ----------- -----  --------- -----------  ---------- ----------- -----------
      TOTAL :                                10800.00   .0       .00       .00       10800.00    10800.00        .00

  015 Form Footings
      RD    10 ES 12SEP95  EF 25SEP95  LS 12SEP95  LF 25SEP95  TF    0

      LABOR                                  12000.00   .0       .00       .00       12000.00    12000.00        .00
                                           ----------- -----  --------- -----------  ---------- ----------- -----------
      TOTAL :                                12000.00   .0       .00       .00       12000.00    12000.00        .00

  020 Pour Concrete Footings
      RD     1 ES 26SEP95  EF 26SEP95  LS  3OCT95  LF  3OCT95  TF    5

      SUB                                     1050.00   .0       .00       .00        1050.00     1050.00        .00
                                           ----------- -----  --------- -----------  ---------- ----------- -----------
      TOTAL :                                 1050.00   .0       .00       .00        1050.00     1050.00        .00

  025 Excavate Piers
      RD     4 ES 12SEP95  EF 15SEP95  LS 20SEP95  LF 25SEP95  TF    6

      LABOR                                   7200.00   .0       .00       .00        7200.00     7200.00        .00
                                           ----------- -----  --------- -----------  ---------- ----------- -----------
      TOTAL :                                 7200.00   .0       .00       .00        7200.00     7200.00        .00

  030 Form Piers
      RD     6 ES 26SEP95  EF  3OCT95  LS 26SEP95  LF  3OCT95  TF    0

      LABOR                                   7200.00   .0       .00       .00        7200.00     7200.00        .00
                                           ----------- -----  --------- -----------  ---------- ----------- -----------
      TOTAL :                                 7200.00   .0       .00       .00        7200.00     7200.00        .00

  035 Pour Concrete Piers
      RD     1 ES  4OCT95  EF  4OCT95  LS  4OCT95  LF  4OCT95  TF    0

      SUB                                     1050.00   .0       .00       .00        1050.00     1050.00        .00
                                           ----------- -----  --------- -----------  ---------- ----------- -----------
      TOTAL :                                 1050.00   .0       .00       .00        1050.00     1050.00        .00

  040 Backfill Footings
      RD     2 ES 27SEP95  EF 28SEP95  LS  4OCT95  LF  5OCT95  TF    5

      LABOR                                   3600.00   .0       .00       .00        3600.00     3600.00        .00
                                           ----------- -----  --------- -----------  ---------- ----------- -----------
      TOTAL :                                 3600.00   .0       .00       .00        3600.00     3600.00        .00

  045 Backfill Piers
      RD     1 ES  5OCT95  EF  5OCT95  LS  5OCT95  LF  5OCT95  TF    0

      LABOR                                   1800.00   .0       .00       .00        1800.00     1800.00        .00
                                           ----------- -----  --------- -----------  ---------- ----------- -----------
      TOTAL :                                 1800.00   .0       .00       .00        1800.00     1800.00        .00

  050 Masonry Block
      RD    10 ES  6OCT95  EF 19OCT95  LS  6OCT95  LF 19OCT95  TF    0

      LABOR                                  14750.00   .0       .00       .00       14750.00    14750.00        .00
                                           ----------- -----  --------- -----------  ---------- ----------- -----------
      TOTAL :                                14750.00   .0       .00       .00       14750.00    14750.00        .00


                                           ----------- -----  --------- -----------  ---------- ----------- -----------
              REPORT TOTALS                  59450.00   .0       .00       .00       59450.00    59450.00        .00
```

Figure 12–1a
Sample project cost report.
This graphic was created using Primavera Project Planner® (P3®), a product of Primavera Systems, Inc.

```
---------------------------------------------------------------------------------------------------------------
Yankee Construction Company                    PRIMAVERA PROJECT PLANNER          Ruth Building

REPORT DATE  29JUL95  RUN NO.   8              COST CONTROL ACTIVITY REPORT       START DATE  4SEP95  FIN DATE 19OCT95
             13:13
CC-10                                                                            DATA DATE  4SEP95   PAGE NO.    1
---------------------------------------------------------------------------------------------------------------
```

ACTIVITY ID	RESOURCE	COST ACCOUNT	ACCOUNT CATEGORY	UNIT MEAS	BUDGET	PCT CMP	ACTUAL TO DATE	ACTUAL THIS PERIOD	ESTIMATE TO COMPLETE	FORECAST	VARIANCE
010	Excavate Footings										
	RD 0										
	LABOR				10800.00	100.0	10800.00	10800.00	1200.00	12000.00	-1200.00
	TOTAL :				10800.00	100.0	10800.00	10800.00	1200.00	12000.00	-1200.00
015	Form Footings										
	RD 0										
	LABOR				13500.00	100.0	13500.00	13500.00	.00	13500.00	.00
	TOTAL :				13500.00	100.0	13500.00	13500.00	.00	13500.00	.00
020	Pour Concrete Footings										
	RD 0										
	SUB				1050.00	100.0	1050.00	1050.00	.00	1050.00	.00
	TOTAL :				1050.00	100.0	1050.00	1050.00	.00	1050.00	.00
025	Excavate Piers										
	RD 0										
	LABOR				7200.00	100.0	7200.00	7200.00	.00	7200.00	.00
	TOTAL :				7200.00	100.0	7200.00	7200.00	.00	7200.00	.00
030	Form Piers										
	RD 3 ES 26SEP95 EF 28SEP95 LS 26SEP95 LF 28SEP95 TF 0										
	LABOR				7200.00	50.0	3600.00	3600.00	3600.00	7200.00	.00
	TOTAL :				7200.00	50.0	3600.00	3600.00	3600.00	7200.00	.00
035	Pour Concrete Piers										
	RD 1 ES 4OCT95 EF 4OCT95 LS 4OCT95 LF 4OCT95 TF 0										
	SUB				1050.00	.0	.00	.00	1050.00	1050.00	.00
	TOTAL :				1050.00	.0	.00	.00	1050.00	1050.00	.00
040	Backfill Footings										
	RD 1 ES 27SEP95 EF 27SEP95 LS 4OCT95 LF 4OCT95 TF 5										
	LABOR				3600.00	50.0	1800.00	1800.00	1800.00	3600.00	.00
	TOTAL :				3600.00	50.0	1800.00	1800.00	1800.00	3600.00	.00
045	Backfill Piers										
	RD 1 ES 5OCT95 EF 5OCT95 LS 5OCT95 LF 5OCT95 TF 0										
	LABOR				1800.00	.0	.00	.00	1800.00	1800.00	.00
	TOTAL :				1800.00	.0	.00	.00	1800.00	1800.00	.00
050	Masonry Block										
	RD 10 ES 6OCT95 EF 19OCT95 LS 6OCT95 LF 19OCT95 TF 0										
	LABOR				14750.00	.0	.00	.00	14750.00	14750.00	.00
	TOTAL :				14750.00	.0	.00	.00	14750.00	14750.00	.00
	REPORT TOTALS				60950.00	62.3	37950.00	37950.00	24200.00	62150.00	-1200.00

Figure 12–1b
Cost report, page 2.

numerous items to control and monitor. The detailed estimate will provide specific direct cost targets such as material quantities, labor rates, and equipment rates and hours, as well as indirect cost elements such as field overhead, contingency, and home office overhead.

At this stage the application of Pareto's Law becomes important. By the use of detailed and summary reports, project managers are able to focus on the elements in need of tight control while they look at less critical elements at a summary level. The report shown in Figure 12–2 illustrates how cost information can be examined and detailed in a table, histogram, and pie chart.

Time Baseline

Through the schedule the project team manages the time and resources required to complete the project. When combined with the estimate, the project's cash flow can be projected. To do this accurately, the schedule must be managed and continually updated to continue to reflect the work that is occurring in the field. As with cost control, the schedule can be used at a summary level or in detail, depending on the level of control required. By the use of bid packages, the WBS, and the CSI format, the necessary levels of control can be established. Detailed control can be achieved by the use of CPM based network schedules, whereas summary control can be provided by the use of bar charts or timetables. Illustrated in Figure 12–3 is the same project described first at the summary level (12–3a) and next (12–3b) at the detailed level.

Most projects will be developed in full detail using a network based CPM schedule with summary reports generated as necessary for reporting purposes. For example, on large projects detailed network schedules will be developed for each of the major systems, which is necessary to coordinate and control the system work. However, as the project is looked at in its entirety, the system work is shown only as a single bar. Most modern day scheduling software allow reports to be generated at different levels of detail as well as to be sorted by area and/or responsibility (see Fig. 12–4).

In developing the schedule for the project it is important not to overconstrain the field activities of the project. As an example, do not specify the specific order in which each drywall partition is to be installed by breaking the work down into specific wall elements. Instead, identify the work as one activity and allow the field superintendent to decide in what order to do the work. Control systems are most effective when the level of control is appropriate for the work and people being controlled. As level of control and corresponding detail increase, the cost of the system also increases. This occurs because of the need to gather, store, and process more information. A high level of control may be necessary on fast paced, highly technical work, or on a project with the need to interface with many different parties. Overcontrolled projects hamper the creativeness of the supervisors and end up

----
Acme Motors   PRIMAVERA PROJECT PLANNER   Plant Expansion and Modernization

REPORT DATE 16MAY94 RUN NO. 610   ----Project Schedule----   START DATE 19JUL93   FIN DATE 13FEB95

BUDGET COST BY DEPARTMENT   DATA DATE 27SEP93   PAGE NO.   1

RESOURCE BUDGETED COST BY DEPARTMENT

	ATM ENG	DES ENG	ELECTRCN	PLUMBER	PRG MGR	PROGRMR	Total
CON	0.00	0.00	200,704.00	175,560.00	0.00	0.00	376,264.00
ENG	21,296.00	157,500.00	0.00	0.00	42,399.00	0.00	221,195.00
ISD	14,570.77	0.00	130,816.00	24,024.00	66,425.10	49,140.00	284,975.87
PCH	1,496.00	0.00	0.00	0.00	5,182.10	0.00	6,678.10
Total	37,362.77	157,500.00	331,520.00	199,584.00	114,006.20	49,140.00	889,112.97

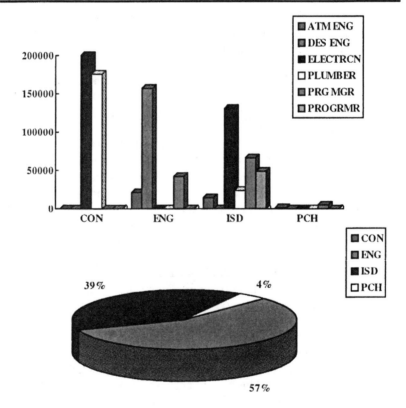

Figure 12–2
Different formatting of cost reports.
This graphic was created using Primavera Project Planner® (P3®), a product of Primavera Systems, Inc.

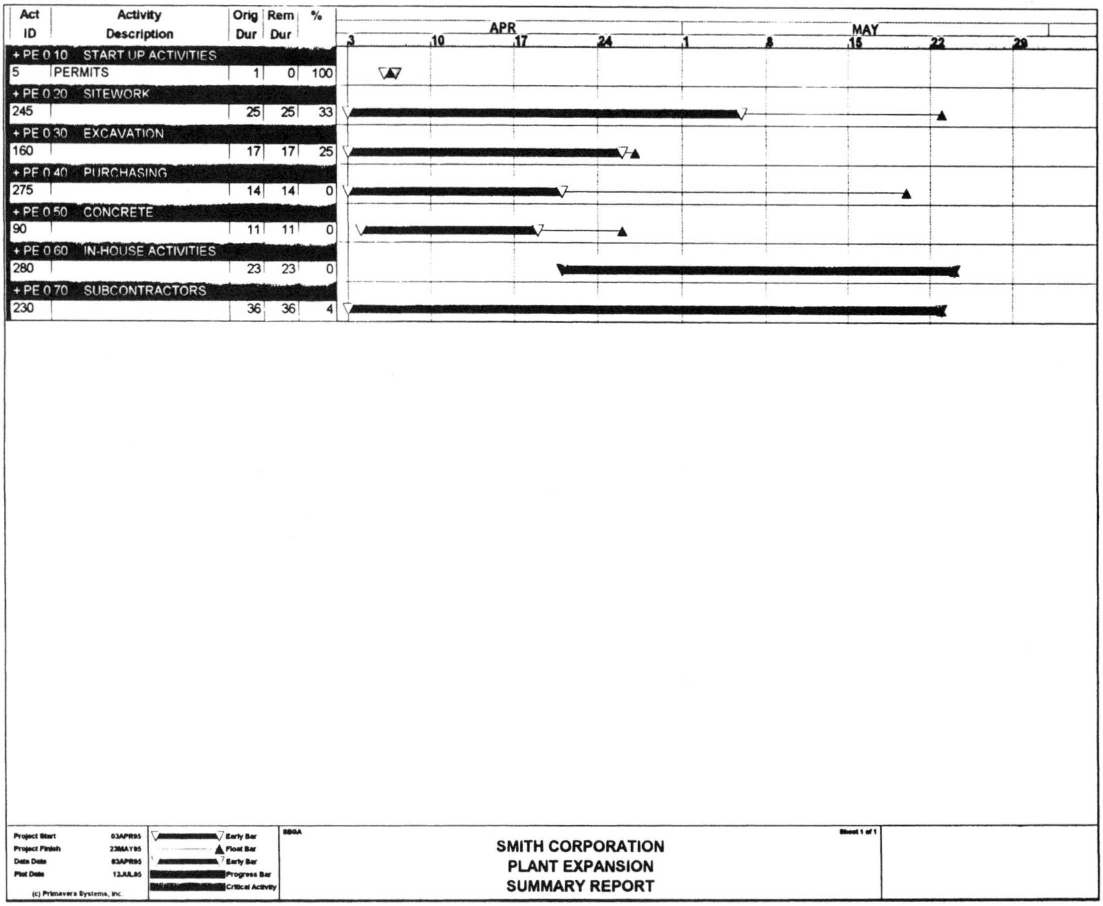

Act ID	Activity Description	Orig Dur	Rem Dur	%
+ PE 0.10	START UP ACTIVITIES			
5	PERMITS	1	0	100
+ PE 0.20	SITEWORK			
245		25	25	33
+ PE 0.30	EXCAVATION			
160		17	17	25
+ PE 0.40	PURCHASING			
275		14	14	0
+ PE 0.50	CONCRETE			
90		11	11	0
+ PE 0.60	IN-HOUSE ACTIVITIES			
280		23	23	0
+ PE 0.70	SUBCONTRACTORS			
230		36	36	4

Project Start	03APR95	
Project Finish	23MAY95	
Data Date	03APR95	
Plot Date	12JUL95	
(c) Primavera Systems, Inc.		

**SMITH CORPORATION
PLANT EXPANSION
SUMMARY REPORT**

Figure 12–3a
Summary level report.
This graphic was created using Primavera Project Planner® (P3®), a product of Primavera Systems, Inc.

wasting time and money. Figure 12–5 provides an example of the criteria which might be used in choosing the level of control to be used vis-à-vis the benefits of early completion.

The baseline schedule identifies the key milestone dates of the project and indicates key material delivery dates. Subcontractor start and finish dates are also shown. These dates are all important control points, as they affect the work of the people involved in negotiating contracts for materials and services. If a material delivery is delayed, it can have a "domino effect" on the follow-on work of the project. Milestone dates, such as the delivery of the first floor for tenant occupancy, are important to monitor, as they may constrain outside users.

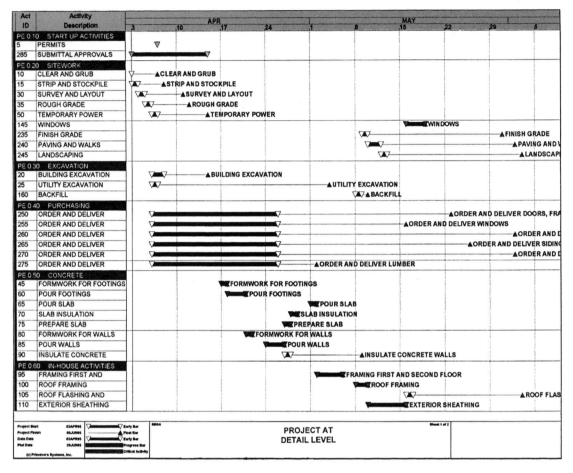

Figure 12–3b
Detailed report.
This graphic was created using Primavera Project Planner® (P3®), a product of Primavera Systems, Inc.

Baseline Summary

The estimate and the schedule are the two primary tools that are used to control the cost and time elements of a project. The level of detail used should be dictated by the degree of control necessary. As the control system is established, it is important that the project manager determine the most time- and cost-sensitive parts of the project and design the control system to focus on these areas. A proper and well thought out breakdown of the work, along with the use of the correct level of summary and detail control, will give the project manager the control tools needed to manage the project in a cost effective manner. The estimate and schedule establish the baseline for the project. As the project proceeds, the actual cost and schedule

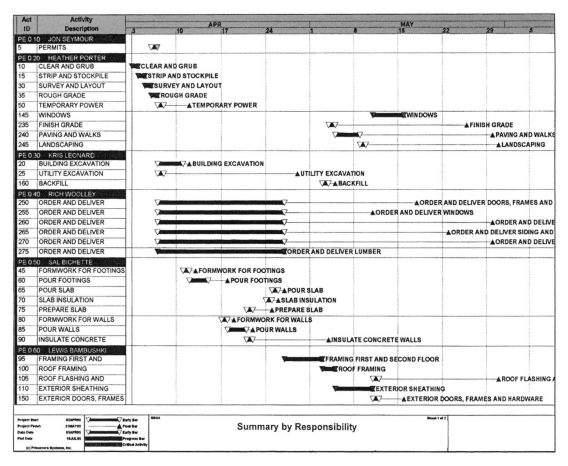

Figure 12–4
Summary report by responsibility.
This graphic was created using Primavera Project Planner® (P3®), a product of Primavera Systems, Inc.

Choosing the Level of Project Control

Ability to Influence Project Duration			
	Low	Medium	High
Benefit of earlier completion			
$5000/mo	Critical Path	Master Project Planning	Cash Flow
$25000/mo	Master Project Planning	Cash Flow	Resource-Loading
$50,000/mo	Cash Flow	Resource-Loading	Resource-Leveling
$100,000+/mo	Resource-Loading	Resource-Leveling	Earned Value& Progress Measurement

Figure 12–5
Fine tuning the project strategy.
Courtesy of Kenneth H. Stowe, P.E., of the George B. H. Macomber Co., Boston, MA

for the project will change for many reasons. The baseline, however, will not change. It will always serve as a measure of how the project was planned at its starting point.

Cost Engineering

For the project team to effectively manage and control the construction process, data used for comparison against the baseline system is first generated in the field. The development of an effective method of gathering and analyzing this data is a key element of maintaining control over the construction process.

People sometimes liken cost engineering to cost accounting, but the two disciplines should not be confused. Both are concerned with costs, but the accountant focuses on historical costs for tax purposes, bill paying, and invoicing of clients, while a cost engineer is concerned with forecasting and trending, and using project information to measure how well a project is doing and what the outcome of the project will be. The two disciplines must work together. The accountant needs project information to properly invoice customers and pay the bills. The project manager uses accounting information for estimating and budgeting purposes.

Cost coding provides the framework by which information is gathered and stored on a project. The coding structure selected determines the level of detail that can be used to study the project and the time and energy that will be required to manage the system. Several structures of accounts are normally used on a project. One is needed to separate costs and operations along the lines of the work breakdown structure. A second structure is necessary to analyze the cost and efficiency of different operations performed on the project. Contractor performance on wall framing, concrete placement, or site clearing is important data to track both to measure performance and to make future estimates. A third structure is needed to manage resources used in requisitioning and reporting. Actual people, equipment, and material costs must be tracked for payroll purposes as well as for invoicing and requisitions. The structure used in this area should be compatible with the company's accounting system.

An example of a project cost code is illustrated below:

Project #	*Area*	*Operation*	*Distribution*	*Cost*
94BR02	02	096600	02	3500

These categories will be discussed in the following paragraphs.

Project

The project number allows the costs to be stored and separated by project type, year, and particular project. In this example the first two fields indicate that this project began in 1994. The third field indicates that the project was won by a competitive bid, and the R indicates that the project is a renovation. The 02 indicates that this is the second project this year for the company.

Area

These two fields allow the stored cost information to be sorted and stored by the area of the particular project. In this case the 02 refers to the second floor on this project. Companies might separate out project information by phase, wings, or other locations on the project.

Operation

In this case the CSI format is used to indicate that the work item being coded is Resilient Tile Flooring. By storing information by operation the company can begin to develop historic costs for different types of materials and operations.

Distribution Code

The 02 in this field indicates that the assigned cost is for material. A 01 might indicate a labor cost, a 03 an equipment cost. This field allows the project manager to assign the cost to vendors, subcontractors, or make other assignments.

Cost

The last entry is the actual cost of the work item. In this case $3,500 was spent to purchase resilient floor tile on this renovation project.

The cost engineering function of a project allows the project manager to input and retrieve information about a project in an organized fashion. It is used to quickly determine the status of a project as well as forecast its future. The development of good historical information for future estimates is also an important component of the cost engineering process.

Progress Evaluation and Control

The estimate and schedule prepared before the project begins establish the baseline for the construction project. Once the project begins, it is important that the actual progress of the project be periodically determined so that any necessary adjustments can be made in a timely fashion. Through the cost-coding system, actual cost information about the project will be collected. Actual progress when combined with actual cost to date allows the productivity on the project to be measured. Productivity information, cost to date, and schedule progress is information that management uses to properly control a project.

The method used to measure work progress is dependent on the type of work. Described next are several different approaches that can be used.

Methods for Measuring Work Progress

Units Completed

When an activity involves the repeated installation or removal of a common piece of work, each repetition involves approximately the same level of effort. In such a case, units completed is used as a straight line measurement. An example might be the installation of floor tile. The job calls for the installation of 2,000 sq.ft. of tile. An evaluation of the work to date shows that 1,500 sq. ft. of tile has been installed, which would indicate that the activity is 75 percent complete.

Incremental Milestone

When an activity involves an operation that will be accomplished in a specific, known sequence, the approximate level of effort necessary to accomplish each milestone (usually based on the number of hours required) is measured. An example might be the installation of a bridge crane in a factory. The milestone completion percentages for the on-site installation might be recorded as follows:

Received and inspected	20%
Crane installed	35%
Alignment completed	50%
Testing completed	90%
Owner accepted	100%

Cost Ratio

This method is used in the case of tasks that occur over a long period of time and are continuous throughout. Project management or a quality assurance program would be examples. These services are budgeted by dollars or work-hours. Percent complete is measured according to the following formula:

$$\text{Percent complete} = \frac{\text{Actual cost or work-hours to date}}{\text{Forecast cost or hours at completion}}$$

Other Methods

The above methods are the most common, but other methods of measuring progress are sometimes used when activities are composed of specific work elements which are of very different cost or work effort. In this case the work may need to be pro-rated to account for the disparity. In some cases—creative work is an example—it is very difficult to determine how long an activity will take and, when in progress, how much has been done. In this case the physical start of the activity can be assigned as 50 percent complete and the completion 100 percent. Lastly, the subjective opinion of the supervisor may be the only measurement option, but this should be done only when other more objective methods are impossible.

The determination of the percent complete of an activity is an important step in the control process. Once the progress of an activity is determined, it is possible to compare the present status of a project with the baseline. This comparison is the best way to determine how the project is proceeding and if any remedial actions need to be taken.

Cost and Schedule Performance

The comparison of actual performance to planned is a critical and recurring step in the control process. Actual performance must be compared to planned performance on an activity-by-activity basis. The level of detail that can be evaluated depends on the level of detail established within the cost-coding system. The frequency of this evaluation depends on the type of project and the level of control required. The greater the level of detail established and the more frequent the reviews, the greater the cost of the control system.

The first step in evaluating the status of a project is to compute the earned value of the activities completed during the reporting period. The **earned value** of an activity, or an account, is computed as follows:

Earned value = Percent complete × Budget for activity

Consider the following example. A framing activity is budgeted to cost $5,000 and consumes 40 work hours. The activity is 40 percent complete as measured by one of the previous methods. Therefore the activity has earned $2,000 and 16 hours.

This process can be accomplished for all the activities computed to date on the project. The dollars and hours earned can also be combined to look at the project as a whole. By combining the earned hours or dollars of all activities to date, the overall percent complete for the project can be calculated as follows:

$$\text{Percent complete} = \frac{\text{Earned work-hours or dollars all accounts}}{\text{Budgeted work-hours or dollars all accounts}}$$

The above earned value calculation computes the work that has been accomplished to date. Budgeted work-hours or dollars is the term used to represent what work has been planned. Actual work-hours or dollars to date will be used to represent what work has been paid for.

Performance will be checked in the following manner:

Schedule performance: Compare budgeted work-hours to earned work-hours. If earned exceeds budgeted, more work has been done than planned.

Budget performance: Compare budgeted work dollars to actual work dollars. If actual exceeds budgeted, more cost has been paid than planned.

Two calculations can be made to further analyze both schedule and cost performance on a project. The schedule or cost **variance** is the first; it is calculated by subtracting the budgeted work-hours from the earned work-hours for the schedule variance, or the actual dollars from the earned dollars for the cost variance.

The second calculation is the performance index, which is calculated by dividing earned work-hours by budgeted work-hours for the schedule index, or earned dollars by budgeted dollars for the cost index. A positive variance (an index of 1.0 or greater) is favorable.

Variance Calculations:

Schedule variance = Earned work-hours – Budgeted work-hours

Cost variance = Earned dollars – Actual dollars paid

Performance Index:

$$\text{Schedule performance index} = \frac{\text{Earned work-hours}}{\text{Budgeted work-hours}}$$

$$\text{Cost performance index} = \frac{\text{Earned dollars}}{\text{Actual dollars paid}}$$

Variance Example

Consider the framing example discussed earlier in this chapter. The activity was budgeted to cost $5,000 and consume 40 work-hours. One carpenter was assigned to the task for a duration of 5 days.

After 3 days (that is, 24 work-hours) the job was determined to be 40 percent complete. To date, $2,550 has been paid out on the activity. What are the schedule and cost variances and indices?

Variances:

$$
\begin{aligned}
\text{Schedule} &= \text{Earned work-hours} - \text{Budgeted work-hours} \\
&= 16 \text{ hours} - 24 \text{ hours} \\
&= -8 \text{ hours}
\end{aligned}
$$

$$\begin{aligned}
\text{Cost} \; &= \; \text{Earned dollars} - \text{Actual dollars} \\
&= \; \$2,000 - \$2,550 \\
&= \; -\$550
\end{aligned}$$

Indices:

$$\begin{aligned}
\text{Schedule performance} \; &= \; \frac{\text{Earned work-hours}}{\text{Budgeted work-hours}} \\[6pt]
&= \; \frac{16}{24} \\[6pt]
&= \; .667 \\[6pt]
\text{Cost performance} \; &= \; \frac{\text{Earned dollars}}{\text{Actual dollars}} \\[6pt]
&= \; \frac{\$2,000}{\$2,550} \\[6pt]
&= \; .784
\end{aligned}$$

The above example indicates a situation where both the cost and schedule performance on the project is not meeting what was planned. The work is taking longer than planned and is costing more. If this type of performance is also occurring on the other project activities, the project might be in serious trouble. What reasons might explain the poor variance and performance indices indicated?

1.   The work being performed is different from the work that was budgeted for. Due to owner or designer changes or differing site conditions, the field crews are doing more work, or work that is of a different type. In this case the schedule and cost budgets must be changed to reflect the actual field work.

2.   The productivity in the field is not as good as planned. This may be due to poor field supervision, the use of the wrong equipment, or a poorly trained work force. Identifying this problem early in the project is essential so that the necessary corrections can be made.

3.   The cost variance and cost performance index is affected by both of the above reasons, as well as by the actual unit prices that are being paid for labor and materials. Due to local market factors, union agreements, or a higher rate of inflation than planned, material and labor prices may be greater than budgeted. Identifying this situation early will allow the project team to adjust the project budget and/or the project's scope.

The calculation of the productivity index for the work can help management better understand which of the above reasons might explain the project performance to date. This index allows for the addition or deletion of work. The productivity index calculation introduces the credit work-hour factor, which measures the actual work being accomplished in the field. The productivity index is calculated as follows:

$$\text{Productivity index} = \frac{\text{Sum of credit work-hours}}{\text{Sum of actual work-hours}}$$

Variance Example Continued

Using the framing example from before, assume that due to an error in estimating the actual work in the field is greater than what was estimated. Actually, 65 hours are required instead of the 40 estimated.

$$\text{Credit work-hours} = 40\% \times 65 \text{ hours}$$

$$= 26 \text{ work-hours}$$

$$\text{Productivity index} = \frac{26 \text{ work-hours}}{24 \text{ work-hours}}$$

$$= 1.08$$

This indicates a productivity level greater than planned.

To be an effective control tool, management should calculate the schedule and cost variances and indices and the productivity index for the project at the end of each reporting period. For the calculations to be useful, it is necessary that the baseline estimate and schedule and the actual cost and performance data be accurate. Remember GIGO (garbage in equals garbage out)! With good data, however, management should be able to quickly identify trouble areas to make timely, intelligent adjustments in the project. The result of this analysis should demonstrate any trends and provide an accurate forecast of the project's future.

Project Documentation

The last step in the project control process is composed of three actions:

1. All actions that occur in a control period need to be documented for administrative and historical reasons.
2. The status of the project and any recommended changes in schedule or budget need to be communicated to all project participants.
3. Through formalized reports the forecasted completion date and cost as well as other critical information such as milestone dates, major purchases, or governmental or regulatory reports must be communicated.

Documentation

By thoroughly documenting project information the project team is able to develop a file of historical information that can be used in a variety of ways. Historical information is necessary in the event of a lawsuit by vendors, suppliers, subcontractors, the owner, or the public. A lawsuit may occur many years into the future, so it is critical that companies establish a formalized system of documenting project events for record purposes. Daily reports, staffing reports, key deliveries, visitors, owner or designer field instructions, tests conducted, activities started or finished, and any unusual occurrences should all be documented for future reference. The original and any revised CPM schedules should be marked up and stored for future use (see again Fig. 7–3).

Accurate actual project information is also necessary for the estimating and scheduling of future projects. The selected system of coding and the effort that goes into inputting and verifying the accuracy of the collected data will dictate the value of this data. Field people must take the time to accurately enter project information as it occurs, and the home office must enter this information into the company data base. Good coordination between the home office and the field will help guarantee an effective data base.

Project Coordination

The level of support from the field in providing data for the cost control system is dependent on the value of the information that is fed back to the field. An action orientated system where analysis is quickly followed up with a recommendation will ensure that information will arrive in the field in time to be implemented. This is a good time for the project managers to applaud positive results or make the necessary changes to get the project back on track.

Instructions back to the field can occur daily, weekly, or monthly depending on the nature of the project. The field people need to know if any changes need to occur in the schedule, the staffing of the project, or planned major purchases or deliveries, and whether or not any overtime or second shifts need to be worked. Feedback as to the approach or equipment being used, good or bad, is essential if productivity improvements are going to be made. It is normal for some activities to be done faster than planned and others to be done slower. Also, some deliveries will arrive early, others late. Work will also be added and deleted. All of these occurrences necessitate that the network schedule be updated and recalculated (see the Project Controls sidebar at the end of this chapter).

The field needs to know if the critical path has changed and what activities are now critical or near critical. Not every activity can be closely monitored, so it is important that the field knows exactly where to focus its attention.

Trending, Forecasting, and Reporting

Looking back at how the project has succeeded, failed, or proceeded is called **trending**. By isolating the different areas of the project over a period of time, project managers are better able to see which decisions have worked and which have not worked and what changes need to be made. Projecting current trends to the future, the project team is better able to forecast future costs and completion dates. The productivity of different trades, material unit prices, labor unit prices, or other indices or variances may be tracked over the project's duration. As each element is better understood, its impact can then be forecasted through the remainder of the project.

It is important for all project participants—owner, designer, construction manager, and contractors—to know with some certainty the ultimate cost and completion date for the project. At the completion of the project everyone will be moving on to other projects, so they each need to be able to predict how long their involvement will be. Each also needs to be able to predict what their financial commitment will be, when money has to be expended, and when they can expect to have money come in to the company. Both time and financial commitments on a project are projected by the use of forecast reports. Calculations such as cost to complete, cost at completion, and projected date of completion are all done by forecasting. In many ways the extension of trending, as described earlier, is the method by which forecasts are made.

A good report should try to include in one format analysis, trending, and forecasting. Some of the questions that should be answered by the report are as follows:

1. How is the project doing (analysis)?

2. Is the production improving (trending)?

3. What is the projected outcome (forecasting)?

Illustrated in the accompanying sidebar provided by Jeffrey Milo are some examples of cost and schedule reports which demonstrate the characteristics of good reporting.

Sidebar

Project Controls

As an electrical subcontractor on a large-scale multimillion dollar sewage treatment facility, it is essential to produce tools that allow the project manager ongoing control over the activities in the project. Through CPM computerized scheduling we are able to produce and continually update a variety of reports that provide timely information. The accompanying arts show a few examples of reports that our company utilizes on projects of this scale. All graphics were created using Primavera Project Planner® (P3®), a product of Primavera Systems, Inc.

Figure A displays a detailed schedule of values for all activities with costs related to them in the schedule. It is used as an application for payment submitted to the client for progress payments within a designated period. The report shown in Figure B allows the scheduler to compare target versus actual/current early, late, or actual dates in a tabular format. It is an essential tool in tracking whether the project is progressing on time.

Figure C is an example of a report that displays a detailed description of each activity's predecessor (PR) and successor (SU) activity. The SS, SF, FS, and FF indicate the relationship within these activities, and the asterisk following the predecessor/successor activity ID identifies it as a driving relationship.

Figures D and E illustrate reports presented in graphical format for ease of understanding. The former displays selected activities in their order of logical occurrence and scheduled time frame. Endpoints display early/actual starts and finishes, while total float arrows represent the relationship between activities. Figure E displays resource allocations spread out over the course of the project. It is very helpful when trying to project labor needs at different stages of the project. As displayed above, it can also illustrate anticipated versus actual resource allocations.

Jeffrey Milo
Project Scheduler
Fischbach & Moore, Inc.

```
X.Y.Z., Inc                         PRIMAVERA PROJECT PLANNER

REPORT DATE 13MAY96  RUN NO.   2         PROJECT CONTROLS          START DATE 09SEP90  FIN DATE 22JAN99*
                     09:58
SPECIFICATION/REQUISITION - DETAIL                                 DATA DATE  13MAY96  PAGE NO.  10
```

ACTIVITY	DESCRIPTION	QUANTITY	U/M	SCHEDULE OF VALUE	PREVIOUS TOTALS	THIS PERIOD	ACTUAL TO DATE	ESTIMATE TO COMPLETE	FORECAST	PCT
4ILFX611	INSTALL LIGHT FIXTURES	40	EA	600	0	0	0	600	600	0
5ILFX611	INSTALL LIGHT FIXTURES	42	EA	800	0	0	0	800	800	0
EILFX611	INSTALL LIGHT FIXTURES	67	EA	2000	0	0	0	2000	2000	0
FILFX611	INSTALL LIGHT FIXTURES	67	EA	2000	0	0	0	2000	2000	0
CILF1611	INSTALL LIGHT FIXTURES	37	EA	1200	0	0	0	1200	1200	0
CILF2611	INSTALL LIGHT FIXTURES	37	EA	1200	0	0	0	1200	1200	0
CILF3611	INSTALL LIGHT FIXTURES	37	EA	1200	0	0	0	1200	1200	0
CILF4611	INSTALL LIGHT FIXTURES	37	EA	1200	0	0	0	1200	1200	0
CILF5611	INSTALL LIGHT FIXTURES	38	EA	1200	0	0	0	1200	1200	0
CILF6611	INSTALL LIGHT FIXTURES	37	EA	1200	0	0	0	1200	1200	0
CILF7611	INSTALL LIGHT FIXTURES	38	EA	1200	0	0	0	1200	1200	0
CILF8611	INSTALL LIGHT FIXTURES	37	EA	1200	0	0	0	1200	1200	0
CILF9611	INSTALL LIGHT FIXTURES	38	EA	1200	0	0	0	1200	1200	0
XILFX611	INSTALL LIGHT FIXTURES	125	EA	6700	0	0	0	6700	6700	0
				300000	8285	0	8285	291715	300000	3
16521	**- ROADWAY LIGHTING**									
016521XX	FAB/DEL - ROADWAY LIGHTING	1	LS	10000	0	0	0	10000	10000	0
IRWLG617	ROADWAY LIGHTING	7	EA	5000	0	0	0	5000	5000	0
				15000	0	0	0	15000	15000	0
16522	**- HIGH MAST LIGHTING**									
016522HM	FAB/DEL - HIGH MAST LIGHTING	1	LS	10000	0	0	0	10000	10000	0
				10000	0	0	0	10000	10000	0
16601	**- LIGHTNING PROTECTION**									
016601LP	FAB/DEL - LIGHTING PROTECTION	1	LS	10000	0	0	0	10000	10000	0
4LPTN301	INSTALL LIGHTNING PROTECTION - ELEC. BLD	1	LS	6000	0	0	0	6000	6000	0
				16000	0	0	0	16000	16000	0
16900	**- ELEC.CONTROLS & MISC.ELEC.EQUIPMENT**									
016900UH	FAB/DEL - UNIT HEATERS	1	LS	2000	0	0	0	2000	2000	0
016900PB	FAB/DEL - PULL BOXES	1	LS	18000	0	0	0	18000	18000	0
016900JB	FAB/DEL - JUNCTION BOXES	1	LS	250000	73997	0	73997	176003	250000	30
016900RC	FAB/DEL - RECEPTACLES	1	LS	10000	0	0	0	10000	10000	0
016900SW	FAB/DEL - SWITCHES	1	LS	20000	0	0	0	20000	20000	0
2SUNH106	SET UNIT HEATERS	1	LS	286	0	0	0	286	286	0
2TUNH306	TEST UNIT HEATERS	1	LS	100	0	0	0	100	100	0
4CPLC38J	CONNECT INSTRUMENTATION PROG. LOGIC CNTR	1	LS	100	0	0	0	100	100	0
5CPLC14J	CONNECT INSTRUMENTATION PROG. LOGIC CNTR	1	LS	100	0	0	0	100	100	0
CCICP226	CONNECT INSTRUMENTATION CONTROL PANELS	1	LS	100	0	0	0	100	100	0
CCICP227	CONNECT INSTRUMENTATION CONTROL PANELS	1	LS	100	0	0	0	100	100	0
CCICP228	CONNECT INSTRUMENTATION CONTROL PANELS	1	LS	100	0	0	0	100	100	0
CCICP229	CONNECT INSTRUMENTATION CONTROL PANELS	1	LS	100	0	0	0	100	100	0
CCICP230	CONNECT INSTRUMENTATION CONTROL PANELS	1	LS	100	0	0	0	100	100	0
CCICP231	CONNECT INSTRUMENTATION CONTROL PANELS	1	LS	100	0	0	0	100	100	0
CCICP232	CONNECT INSTRUMENTATION CONTROL PANELS	1	LS	100	0	0	0	100	100	0
CCICP233	CONNECT INSTRUMENTATION CONTROL PANELS	1	LS	100	0	0	0	100	100	0
CCICP234	CONNECT INSTRUMENTATION CONTROL PANELS	1	LS	100	0	0	0	100	100	0
				301486	73997	0	73997	227489	301486	25
16915	**- ELECTRIC HEAT TRACING**									
016915HT	FAB/DEL - HEAT TRACING & CONTROLS	1	LS	30000	0	0	0	30000	30000	0
EHTCA501	INSTALL HEAT TRACING & CONTROLS - BATTER	2000	LF	25000	0	0	0	25000	25000	0
FHTCB501	INSTALL HEAT TRACING & CONTROLS - BATTER	2000	LF	25000	0	0	0	25000	25000	0
				80000	0	0	0	80000	80000	0
16920	**- 600-VOLT MOTOR CONTROL CENTERS**									
016920MC	FAB/DEL - MCC'S	1	LS	300000	33000	0	33000	267000	300000	11
016920MS	FAB/DEL - MANUAL MOTOR STARTER	1	LS	4000	0	0	0	4000	4000	0
1SMCC201	SET MOTOR CONTROL CENTERS	1	LS	6240	0	0	0	6240	6240	0
2SMCC201	SET MOTOR CONTROL CENTERS	1	LS	6240	0	0	0	6240	6240	0
4SMCC201	SET MOTOR CONTROL CENTERS	1	LS	6680	6680	0	6680	0	6680	100
5SMCC201	SET MOTOR CONTROL CENTERS	1	LS	6640	0	0	0	6640	6640	0
				329800	39680	0	39680	290120	329800	12
				15621500	8150859	0	8150859	7470641	15621500	53

Figure A
Payment requisition report.

```
----------------------------------------------------------------------------------------------------
X.Y.Z., Inc.                          PRIMAVERA PROJECT PLANNER

REPORT DATE 13MAY96  RUN NO.    4           PROJECT CONTROLS            START DATE 09SEP90  FIN DATE 22JAN99*
                    11:09
PRELIMINARY - ORIGINAL VS CURRENT ES DATES                             DATA DATE  13MAY96  PAGE NO.    8
----------------------------------------------------------------------------------------------------
```

ACTIVITY	DESCRIPTION	ORIGINAL DURATION	TARGET EARLY START	TARGET EARLY FINISH	REVISED EARLY START	REVISED EARLY FINISH	REVISED TF
UNDERGROUND DISTRIBUTION RACEWAY SYSTEM							
IDBWEAST	MV/LV DUCTBANK (E4829.94)	40	29APR94	24JUN94	18JUL94A	29JUN95A	
IDBNORTH	MV/LV DUCTBANKS - NORTH	1	06NOV95	20NOV95	05APR96	05APR96	428
IDBWEXXX	MV/LV DUCTBANKS - W/E	1	22SEP95	03NOV95	05APR96	05APR96	243
IDBEB869	MV/LV DUCTBANKS - ELECTRICAL BLDG #4/5	1	15JAN96	09FEB96	05APR96	05APR96	428
ILMDB/MH	MV/LV CONCRETE DUCTBANKS/MANHOLES	1	22SEP95	19DEC95	05APR96	05APR96	428
IMVLV241	MV/LV DUCTBANKS - CP241 HYDRO PLANT	1	06NOV95	05DEC95	05APR96	05APR96	428
GROUNDING							
FUGBB504	UNDERSLAB GROUNDING - BATTERY F	80	09JAN95	02MAY95	19SEP95A	03JUN96	209
EUGAN504	UNDERSLAB GROUNDING - BATTERY E (NORTH)	30			05APR96	17MAY96	56
EUGAS504	UNDERSLAB GROUNDING - BATTERY E (SOUTH)	50	02SEP94	29DEC94	07JUL95A	21JUL95A	
CUGMG504	UNDERSLAB GROUNDING - MAIN CORRIDORE	5	30AUG95	06SEP95	09MAY95A	05APR96	96
CUGMN504	UNDERSLAB GROUNDING - MAIN CORRIDORENORTH	5			22APR96	26APR96	76
CUGMS504	UNDERSLAB GROUNDING - MAIN CORRIDORE	5	27MAY94	03JUN94	08SEP94A	21JUL95A	
LIGHTING							
XILFX611	INSTALL LIGHT FIXTURES	10	26JUN95	10JUL95	10JUL96	23JUL96	67
FILFX611	INSTALL LIGHT FIXTURES	10	04AUG95	17AUG95	27JUN96	11JUL96	147
EILFX611	INSTALL LIGHT FIXTURES	10	24MAY95	07JUN95	04JUN96	17JUN96	22
CILF1611	INSTALL LIGHT FIXTURES	10	23AUG95	29AUG95	19JUL96	01AUG96	17
CILF2611	INSTALL LIGHT FIXTURES	10	09AUG95	15AUG95	08JUL96	19JUL96	29
CILF3611	INSTALL LIGHT FIXTURES	10	19JUL95	25JUL95	25JUN96	09JUL96	36
CILF4611	INSTALL LIGHT FIXTURES	10	20JUN95	26JUN95	23MAY96	06JUN96	49
CILF5611	INSTALL LIGHT FIXTURES	10	24FEB95	01MAR95	01JUL96	15JUL96	23
CILF6611	INSTALL LIGHT FIXTURES	10	08FEB95	14FEB95	01JUL96	15JUL96	18
CILF7611	INSTALL LIGHT FIXTURES	10	12JAN95	17JAN95	05JUN96	18JUN96	31
CILF8611	INSTALL LIGHT FIXTURES	10	19DEC94	23DEC94	23MAY96	06JUN96	29
CILF9611	INSTALL LIGHT FIXTURES	10	02DEC94	02DEC94	20MAY96	03JUN96	32
4ILFX611	INSTALL LIGHT FIXTURES	5	16MAR95	22MAR95	09APR96	16APR96	40
5ILFX611	INSTALL LIGHT FIXTURES	5	18SEP95	22SEP95	15MAY96	21MAY96	15
1ILFX611	INSTALL LIGHT FIXTURES	5	04JAN95	10JAN95	22APR96	26APR96	82
2ILFX611	INSTALL LIGHT FIXTURES	5	23AUG95	29AUG95	02OCT96	08OCT96	107
ROADWAY LIGHTING							
IRWLG615	ROADWAY LIGHTING	1	26JAN96	08MAR96	05APR96	05APR96	373
LIGHTNING PROTECTION							
4LPTN621	INSTALL LIGHTNING PROTECTION - ELEC. BLDG	10	09AUG95	06SEP95	20SEP95A	14JUN96	98
5LPTN621	INSTALL LIGHTNING PROTECTION - ELEC. BLDG	10			22JUL96	02AUG96	64
ELECTRICAL CONTROLS & MISC EQUIPMENT							
CCIC1705	CONNECT INSTRUMENTATION CONTROL PANELS	10	20JUN95	26JUN95	19JUL96	01AUG96	17
CCIC2705	CONNECT INSTRUMENTATION CONTROL PANELS	10	26APR95	01MAY95	28JUN96	12JUL96	34
CCIC3705	CONNECT INSTRUMENTATION CONTROL PANELS	10	04APR95	10APR95	07JUN96	20JUN96	48
CCIC4705	CONNECT INSTRUMENTATION CONTROL PANELS	10	28FEB95	06MAR95	23MAY96	06JUN96	54
CCIC5705	CONNECT INSTRUMENTATION CONTROL PANELS	10	06FEB95	10FEB95	17JUL96	30JUL96	12
CCIC6705	CONNECT INSTRUMENTATION CONTROL PANELS	10	22DEC94	28DEC94	02JUL96	16JUL96	12
CCIC7705	CONNECT INSTRUMENTATION CONTROL PANELS	10	16NOV94	21NOV94	18JUN96	01JUL96	12
CCIC8705	CONNECT INSTRUMENTATION CONTROL PANELS	10	11OCT94	14OCT94	04JUN96	17JUN96	12
CCIC9705	CONNECT INSTRUMENTATION CONTROL PANELS	10	19SEP94	19SEP94	20MAY96	03JUN96	12
4CPLC705	CONNECT INSTRUMENTATION PROG. LOGIC CNTRL	5	26APR95	02MAY95	04JUN96	10JUN96	2
5CPLC705	CONNECT INSTRUMENTATION PROG. LOGIC CNTRL	5	18SEP95	22SEP95	04JUN96	10JUN96	2
2SUNH710	SET UNIT HEATERS	2	25OCT95	26OCT95	30SEP96	01OCT96	112
2TUNH715	TEST UNIT HEATERS	10	05SEP95	11SEP95	13JAN97	24JAN97	104
ELECTRIC HEAT TRACING							
FHTCB501	INSTALL HEAT TRACING & CONTROLS - BATTERY	80	18SEP95	12JAN96	13FEB97	09JUN97	4
EHTCA501	INSTALL HEAT TRACING & CONTROLS - BATTERY	30	13JAN95	08MAY95	18JUN96	30JUL96	19
600-VOLT MOTOR CONTROL CENTERS							
4SMCC101	SET MOTOR CONTROL CENTERS	10	13FEB95	27MAR95	20MAR96A	22MAR96A	
4SMCC301	TEST MOTOR CONTROL CENTER -	5			21MAY96	28MAY96	6
5SMCC101	SET MOTOR CONTROL CENTERS	10	12JUL95	08AUG95	01MAY96	14MAY96	0

Figure B
Target vs. actual report.

ACTIVITY ID	ORIG DUR	REM DUR	CAL	%	CODE	ACTIVITY DESCRIPTION	EARLY START	EARLY FINISH	LATE START	LATE FINISH	TOTAL FLOAT
..CEQCN209*	10	10	1	0	PR FF 0	ELECTRICAL CONNECTIONS (LOCAL)	20MAY96	3JUN96	5JUL96	18JUL96	32
CILFX611	10	10	1	0		INSTALL LIGHT FIXTURES	20MAY96	3JUN96	5JUL96	18JUL96	32
..CTEEQ311*	5	5	1	0	SU	ELECTRICAL TESTING (LOCAL)	4JUN96	10JUN96	19JUL96	25JUL96	32
..CCICP226*	10	10	1	0	PR	CONNECT INSTRUMENTATION CONTROL PANELS	20MAY96	3JUN96	6JUN96	19JUN96	12
..CEQCN209	10	10	1	0	PR	ELECTRICAL CONNECTIONS (LOCAL)	20MAY96	3JUN96	5JUL96	18JUL96	32
..CILFX611*	10	10	1	0	PR	INSTALL LIGHT FIXTURES	20MAY96	3JUN96	5JUL96	18JUL96	32
CTEEQ311	5	5	1	0		ELECTRICAL TESTING (LOCAL)	4JUN96	10JUN96	19JUL96	25JUL96	32
4A07B901	30	0	1	100		MEDIUM VOLTAGE CONDUIT - A09/B09	20SEP95A	16FEB96A			
..4A07B509	15	15	1	0	SU	PULL MEDIUM VOLTAGE CABLE - 15KVA (A09/B09)	11APR96	2MAY96	17APR96	7MAY96	3
..5A07B901*	30	15	1	50	SU	MEDIUM VOLTAGE CONDUIT - A09/B09	20FEB96A	3MAY96		3MAY96	0
..4A07B901*	30	0	1	100	PR	MEDIUM VOLTAGE CONDUIT - A09/B09	20SEP95A	16FEB96A			
4SLBS109	2	0	1	100	PR	SET 15KV LOAD BREAKER SWITCHES	28NOV95A	25JAN96A			
4A07B509	15	15	1	0		PULL MEDIUM VOLTAGE CABLE - 15KVA (A09/B09)	11APR96	2MAY96	17APR96	7MAY96	3
..4STRF106	10	0	1	100	SU SS 10	SET KVA TRANSFORMERS	21NOV95A	25JAN96A			
..4TIEIN01	5	5	1	0	SU	CIRCUIT A09/B09 - MV CABLE TIE-IN	3MAY96	9MAY96	30MAY96	5JUN96	18
..4TLBS309*	3	3	1	0	SU	TEST 15KV LOAD BREAKER SWITCHES	3MAY96	7MAY96	13MAY96	15MAY96	6
4EMCD403	20	0	1	100		EMBEDDED CONDUITS	22AUG95A	19SEP95A			
..4RIES410*	5	2	1	55	SU	ROUGH IN ELECTRICAL SYSTEMS	25MAR96A	8APR96		7MAY96	20
..016601LP*	120	40	1	67	PR	FAB/DEL - LIGHTING PROTECTION	20SEP95A	3JUN96		21OCT96	97
4LPTN515	10	9	1	10		INSTALL LIGHTNING PROTECTION - ELEC. BLDG 4	20SEP95A	14JUN96		4NOV96	98
..016332ST	100	0	1	100	PR	FAB/DEL - SUBSTATION TRANSFORMERS	4APR95A	4APR95A			
..4A07B509*	15	15	1	0	PR SS 10	PULL MEDIUM VOLTAGE CABLE - 15KVA (A09/B09)	11APR96	2MAY96	17APR96	7MAY96	3
4STRF106	10	0	1	100		SET KVA TRANSFORMERS	21NOV95A	25JAN96A			
..4SSWG102*	5	5	1	0	SU	SET SWITCHGEAR	26APR96	2MAY96	1MAY96	7MAY96	3
..4TTRF106	4	4	1	0	SU	TEST 1000 KVA TRANSFORMERS	8MAY96	13MAY96	16MAY96	21MAY96	6
4SLBS109	2	0	1	100		SET 15KV LOAD BREAKER SWITCHES	28NOV95A	25JAN96A			
..4A07B509*	15	15	1	0	SU	PULL MEDIUM VOLTAGE CABLE - 15KVA (A09/B09)	11APR96	2MAY96	17APR96	7MAY96	3
4EXCD404	5	2	1	60		EXPOSED CONDUIT	23JAN96A	8APR96		5JUN96	40
..4ILFX611*	5	5	1	0	SU	INSTALL LIGHT FIXTURES	9APR96	16APR96	6JUN96	12JUN96	40
..016920MC	160	0	1	100	PR SS 60	FAB/DEL - MCC'S	6JUL94A	20MAR96A			
..4SSWG102*	5	5	1	0	SU	SET SWITCHGEAR	26APR96	2MAY96	1MAY96	7MAY96	3
4SMCC103	10	0	1	100		SET MOTOR CONTROL CENTERS	20MAR96A	22MAR96A			
..4BDCT105	20	20	1	0	SU SS 5	INSTALL BUS DUCT - ELEC BLDG. #4	22APR96	17MAY96	8MAY96	5JUN96	12
..4CVFD107	20	20	1	0	SU	SET VFD CONTROL PANELS	3MAY96	31MAY96	19AUG96	16SEP96	74
..4SLVT104	2	2	1	0	SU	SET LOW VOLTAGE TRANSFORMERS	3MAY96	6MAY96	4JUN96	5JUN96	21
..4SMCC303*	5	5	1	0	SU	TEST MOTOR CONTROL CENTER - ELEC. BLDG. #4	21MAY96	28MAY96	30MAY96	5JUN96	6
..4EMCD403*	20	0	1	100	PR	EMBEDDED CONDUITS	22AUG95A	19SEP95A			
4RIES410	5	2	1	55		ROUGH IN ELECTRICAL SYSTEMS	25MAR96A	8APR96		7MAY96	20
..4WELS406*	15	15	1	0	SU	WIRE ELECTRICAL - ELECTRICAL BLDG #4	9APR96	30APR96	8MAY96	29MAY96	20
..4EXCD403*	5	2	1	60	PR	EXPOSED CONDUIT	23JAN96A	8APR96		5JUN96	40
4ILFX611	5	5	1	0		INSTALL LIGHT FIXTURES	9APR96	16APR96	6JUN96	12JUN96	40
..4ENGZ907*	5	5	1	0	SU	ENERGIZE ELECTRICAL BLDG #4	12JUN96	18JUN96	13JUN96	19JUN96	1
..4RIES410	5	2	1	55	PR	ROUGH IN ELECTRICAL SYSTEMS	25MAR96A	8APR96		7MAY96	20
4WELS406	15	15	1	0		WIRE ELECTRICAL - ELECTRICAL BLDG #4	9APR96	30APR96	8MAY96	29MAY96	20
..4EQCN208*	5	5	1	0	SU	CONNECT ELECTRICAL EQUIPMENT - BATTERY E	1MAY96	7MAY96	30MAY96	5JUN96	20
..4EQCN209	20	20	1	0	SU	ELECTRICAL CONNECTIONS BLDG. #4 (BATTERY F)	1MAY96	29MAY96	6MAR97	2APR97	212
..016112BD*	120	10	1	92	PR	FAB/DEL - BUS DUCT	9AUG95A	19APR96		7MAY96	12
..4SMCC103	10	0	1	100	PR SS 5	SET MOTOR CONTROL CENTERS	20MAR96A	22MAR96A			
4BDCT105	20	20	1	0		INSTALL BUS DUCT - ELEC BLDG. #4	22APR96	17MAY96	8MAY96	5JUN96	12
..4BDCT305*	5	5	1	0	SU	TEST BUS DUCTS - ELECTRICAL BLDG. #4	29MAY96	4JUN96	6JUN96	12JUN96	6
..416320SG*	125	10	1	92	PR	FAB/DEL - SWITCHGEAR	7JUL95A	19APR96		30APR96	7
..4STRF106	10	0	1	100	PR	SET KVA TRANSFORMERS	21NOV95A	25JAN96A			
4SSWG102	5	5	1	0		SET SWITCHGEAR	26APR96	2MAY96	1MAY96	7MAY96	3

Figure C
Predecessor and successor report.

Activity ID	Activity Description	Orig Dur	Rem Dur	%
ELECTRICAL BUILDING #4				
4EMCD403	EMBEDDED CONDUITS	20	0	100
4LPTN405	INSTALL LIGHTNING PROTECTION - ELEC. BLDG 8	10	9	10
4A07B901	MEDIUM VOLTAGE CONDUIT - A09/B09	30	0	100
4STRF106	SET KVA TRANSFORMERS	10	0	100
4SLBS109	SET 15KV LOAD BREAKER SWITCHES	2	0	100
4EXCD405	EXPOSED CONDUIT	5	2	60
4SMCC103	SET MOTOR CONTROL CENTERS	10	0	100
4RIES410	ROUGH IN ELECTRICAL SYSTEMS	5	2	55
4WELS406	WIRE ELECTRICAL - ELECTRICAL BLDG #4	15	15	0
4ILFX611	INSTALL LIGHT FIXTURES	5	5	0
4A07B509	PULL MEDIUM VOLTAGE CABLE - 15KVA (A09/B09)	15	15	0
4BDCT105	INSTALL BUS DUCT - ELEC BLDG. #4	20	20	0
4SSWG102	SET SWITCHGEAR	5	5	0
4EQCN208	CONNECT ELECTRICAL EQUIPMENT - BATTERY E	5	5	0
4EQCN209	ELECTRICAL CONNECTIONS BLDG. #4 (BATTERY F)	20	20	0
4TLBS319	TEST 15KV LOAD BREAKER SWITCHES	3	3	0
4TIEIN01	CIRCUIT A07/B07 - MV CABLE TIE-IN	5	5	0
4SLVT104	SET LOW VOLTAGE TRANSFORMERS	2	2	0
4CVFD107	SET VFD CONTROL PANELS	20	20	0
4TLVT308	TEST LOW VOLTAGE MANUAL TRANSFER SWITCHES	2	2	0
4SSSP110	SET SWITCHES/STARTERS/PANELS	5	5	0
4TRF306	TEST 1000 KVA TRANSFORMERS	4	4	0
4TIEIN02	CIRCUIT A07/B07 - CHECKOUT/TESTING (BLDG #8)	5	5	0

Sheet 1A of 3B

X.Y.Z., Inc.
Project Controls
Classic Schedule Layout

Project Start	09SEP90	Early Bar
Project Finish	29DEC97	Float Bar
Data Date	05APR96	Progress Bar
Plot Date	21MAY96	Critical Activity

1234

© Primavera Systems, Inc.

Figure D
Time-scaled logic report.

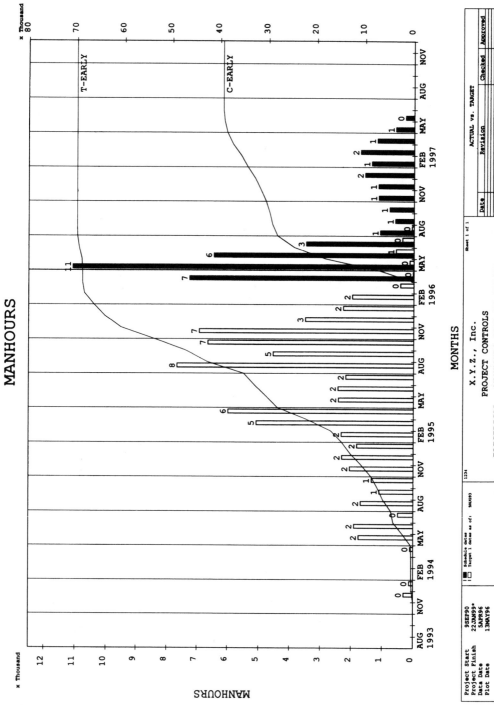

Figure E
Resource loading graph.

Conclusion

A good control process is action-oriented. Project managers must know what their goals are and then communicate them to all the key participants, measure results, and take corrective action as necessary. In construction, the goals of a project are established by the estimate and schedule, which in turn establish the cost and schedule standards for the project. These goals are transferred to the field through specific tasks, each of which has budget and schedule constraints. Field performance is periodically measured with the actual results compared to the set standards. The level of detail that is measured is determined by the cost coding system that was adopted for the project. As the level of detail increases, the cost of managing the control process increases.

Performance is calculated by computing cost and schedule variances and performance indices. Productivity performance is also measured. Managers use these calculations to analyze the project's performance and make decisions as to any necessary changes which need to be made on the project. Actual performance data and any other information about the project needs to be documented and stored. Management decisions need to be communicated promptly to all key project participants. A timely response by management provides feedback to the field and provides ample opportunity for the field to implement any recommendations. The last control responsibility is for management to continually report on the progress of the project. Reports should be timely, should indicate key variances between budgets and actuals, should project trends, and should forecast the project's completion cost and date.

Chapter Review Questions

1. A cost variance shows whether work performed costs more or less than budgeted.

   ___ T   ___ F

2. Cost control is a form of cost accounting.

   ___ T   ___ F

3. A project control system should isolate and control in detail those elements with the greatest impact on final cost.

   ___ T   ___ F

4. The best project cost coding systems are designed by accountants for historical data collection.

   ___ T   ___ F

5. Reporting data needs to consider productivity, quantities budgeted and utilized, and labor budgeted and expended to properly analyze a field problem.

   ___ T   ___ F

6. The budget baselines of a project are generated through:
   a. Scheduling
   b. Estimating
   c. Random guesses
   d. Historical data

7. Cost control should be approached as an application of Pareto's Law, which states:
   a. 50 percent of the outcome of a project is determined by 50 percent of the included elements
   b. 80 percent of the outcome of a project is determined by 20 percent of the included elements
   c. 25 percent of the outcome of a project is determined by 75 percent of the included elements

8. Which of the below listed methods of progress evaluation should only be used as a last resort?
   a. Units completed
   b. Incremental milestone
   c. Cost ratio
   d. Supervisor opinion

9. Which of the below listed reasons would explain a negative schedule variance?
   a. An unrealistic schedule baseline
   b. Poor field productivity
   c. Added work requirements
   d. Poor field supervision
   e. All of the above

10. The productivity index calculation is a necessary calculation since it allows for:
    a. Poor weather
    b. The addition or deletion of work
    c. An increase in labor unit prices
    d. Pareto's Law

Exercises

1. Gather some examples of job cost reports from several different design or construction companies. Evaluate how well the reports meet the principal objectives of project reporting.

2. Investigate how local companies gather field information for home office use. What systems are used to utilize this information for future projects?

Sources of Additional Information

Adrien, James J. *Construction Productivity Improvement*. New York: Elsevier, 1987.

Mueller, Frederick Wm. *Integrated Cost and Scheduling Control*. New York: Van Nostrand Reinhold Co., Inc., 1986.

Parker, Henry W., and Clarkson H. Oglesby. *Methods Improvement for Construction Managers*. New York, McGraw-Hill, 1972.

Pierce, David R. *Project Planning and Control for Construction*. Kingston, MA: R. S. Means Co., Inc., 1988.

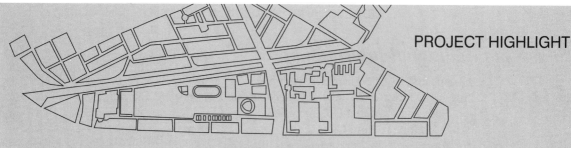

MIT Renovation of Building 16 and 56

SECTION FOUR

Project Control

Nancy E. Joyce

Developing tools to control the design and construction of Building 16 and 56 began very early in the design process. During the design process, costs can escalate on the project through a variety of means. User groups can request enhancements; field conditions can dictate added scope; designers can specify more expensive methods or materials than anticipated. These are all natural phenomena occurring during the development of the design. The team, however, needs to develop methods to identify and control these occurrences so that the scope and costs are known early and the owner can make informed decisions.

With so many different user groups in the program, everyone recognized the potential for costs to be driven up through customization for each group. Before the design team began programming and schematic design, we developed a method of organizing the space through the formulation of standards. The standards included a core facility on each floor, standard layouts for laboratories, and a standardized approach to the administrative areas. The core facility contained rest rooms, conference rooms, student lounges, and shared researcher laboratory-support space. These spaces were placed in the center of the two buildings around the existing stairs and elevators and acted as a strong organizing element in the building. In the laboratories we developed standard layouts within the existing window bays. These layouts were configured with modular casework units and standardized zones of activities to facilitate future changes. By modularizing the laboratory casework, we were also able to keep the costs for custom casework to a minimum. We also standardized the approach to the administrative areas. To maximize flexibility, we kept all offices the same size and provided more open areas with modular and moveable partitions.

The creation and maintenance of standards provided the Institute with multiple advantages. Besides long term flexibility, it also cut down on first costs that would have been spent creating custom spaces, minimized future costs when changes occurred, and created a simpler building to operate and maintain.

To minimize the impact of field conditions, the owner directed the team to investigate the existing conditions in the building, identify what systems could be retained and what costs would be associated with their reuse. The early feasibility studies showed that the equipment in the buildings was at the end of its useful life. We assumed that these fixtures would be replaced and concentrated instead on the HVAC, plumbing, and electrical distribution systems. The electrical riser system at first look seemed capable of being reused because it was in good condition. However, because of its age the parts were hard to get and the placement of it through the building made it difficult to work around. When we factored in the repair costs associated with keeping the system, the gap between buying new and retaining old narrowed, and the owner decided to go with a new system. We also tested the plumbing and heating pipes to determine the extent of their corrosion. The results showed that we could retain some of these. The perimeter radiation piping could be retained with new covers and new controls. The piping risers had some corrosion and since they would be hard to access for replacement in the future, it was decided to replace these. The horizontal runs, however, were retained since they would be easier to access. The fume hood exhaust ducts in Building 56 were determined to be reusable but would have to be cleaned. The elevators were retained, although an extensive repair and replacement of parts as well as a code upgrade was included as part of the scope.

The early investigation of the buildings identified reusable components. The analysis of these components identified the scope required to refurbish them for reuse. This information gave the team the tools for making realistic decisions about what to retain and allowed for early inclusion of these costs in the estimate.

Another area where costs can creep into a project is through designer specified standards. Design firms have their own particular design standards that they have developed over the years through experience of what materials and methods have proven successful. Owners also have standards of what materials and methods work best for operation and maintenance. These standards are not always the same. MIT publishes a guide for designers that sets out in CSI format what level of performance is needed from their buildings. In the renovation of Buildings 16 and 56 these guides acted as a baseline from which a series of discussions took place between the project team and the MIT in-house operations group. The guidelines ensured that MIT would be handed over an easily maintainable building, at a specified level of quality. Because the guidelines were performance oriented and not product oriented they easily incorporated changes. However, it did put the responsibility in the architects' hands to identify where and why changes were being recommended. Any costs associated with these changes could then also be identified so that the owner could make an informed choice.

To keep track of all these areas where costs could escalate, the team developed a scope log that recorded any changes made after schematic design (see Figs. A–D). The estimate at schematic design incorporated the broad scope of the project that was bracketed by the formulation of the user group standards, the analysis of reusable components, and the MIT guidelines. During design development, as changes occurred, they were recorded on the log and priced out. We also used the log for cost reduction ideas that the team developed. All items stayed on the spreadsheet log as a record of decisions. With its cost plus and minuses, this log helped the team to make decisions about changes to the project before the changes were formalized on the drawings. This process saved time and money in the redesign effort that would have been necessary if the changes were not identified until the full cost estimating exercise at the end of design development or at bid time.

Beacon Construction Company					22-May-95

Massachusetts Institute of Technology
Buildings 16 and 56 Renovations
Potential Cost Impacts
ACCEPTED

Major System	Item of Work	Phase	Description	Comments	Estimate
Interior Construction					
IC03	Ceilings	1&2	Delete ceiling veneer plaster		($3,600)
IC04	Program	1	Price Arch and Mech, Elect changes at 8th	now includes item E-12	$200,500
IC04	Program	2	Price Arch and Mech, Elect changes at 8th		$118,800
IC05	Ceilings	1&2	Delete open grid ceilings at flrs 2-8		($156,000)
IC08	Laboratory casework	1&2	Additional casework in research labs		($112,045)
HVAC					
H01	Exhaust	1	Gang exhaust into plenums in penthouse		($250,000)
H01	Exhaust	2	Gang exhaust into plenums in penthouse		($30,000)
H02	Exhaust	1&2	Locate trunk at south corridor side		($3,000)
H03	Distribution	1	Substitute radiation at walls and fan coils at ceilings		$533,500
Plumbing					
P05	Lab grade water	1&2	Alternate piping in lieu of Kynar	Accepted 80%	($200,000)
P07	Emergency showers	1&2	Emergency showers grouped per valve		($28,000)
P09	Water pressure	1&2	Dupl.vs.triplex water pressure booster		($4,000)
Fire Protection					
FP02	Sprinklers	1&2	Reduce shaft sprinklering	Formerly P02	($175,000)
FP03	Sprinklers	1&2	Fire main conn to Bldg 68 pump	Formerly P03	$55,000
Electrical					
E01	Handicap Elevator	2	Reduce new elevator to 6 stops		($50,000)
E04	Emergency Power	1&2	Delete elevator emergency power		($14,750)
E09	Distribution	1&2	Provide AC cable in lieu of conduit		($74,000)
E10	Lightning Protection	1&2	Delete lightning protection system		($35,000)
Equipment					
EQ02	Autoclave/Glasswash	1&2	Substitute Gentinge autoclave #6912AR1		($83,500)
EQ03	Autoclave/Glasswash	1&2	Substitute Lancer glasswash		($65,000)
EQ04	Autoclave/Glasswash	1&2	Substitute smaller Getinge autoclave #6612AR1	add'l (-13,250/ea) applied to EQ02	T B D
TOTAL					($376,095)

Figure A
Scope log—potential cost impacts, accepted.

Beacon Construction Company 22-May-95

Massachusetts Institute of Technology
Buildings 16 and 56 Renovations
Potential Cost Impacts
PENDING

Major System	Item of Work	Phase	Description	Comments	Estimate
Structural					
S01	Spread Footings	2	Alternate to PIFs	Geotech input required	($80,000)
S03	Roof Screen	2	Reduce Cost of Roof Screen	EAI to study as part of DD	
S04	Stack Support	1&2	Reduce cost of stack supports	Ganging scheme will have impact	($25,000)
S05	Exhaust Stacks	1&2	Cluster stacks at roof	EAI to study as part of DD	($15,000)
S06	Service Distribution	1&2	Combine services in shafts	Part of systems design	($30,000)
S07	Shafts	1	Fit shafts between joists	Part of systems design	
S08	Shafts	2	Fit shafts between beam faces	Part of systems design	
S09	Foundation	2	Excavate area between foundation and bldg.	Included w/ SD estimate	---
Exterior Construction					
EC05-a	Siding	2	3 Alts for materials @ the 8th floor-North side	option a - Curtainwall	$5,400
EC05-b	Siding	2	3 Alts for materials @ the 8th floor-North side	option b - Panels w/ ribbon windows	($500)
EC05-c	Siding	2	3 Alts for materials @ the 8th floor-North side	option c - Panels w/ punched windows	($2,550)
Interior Construction					
IC01	Corridors	1&2	Simplify glazing & metal panel design	EAI to study as part of DD	($25,000)
IC02	Partitions	1&2	Reduce acoustic rating in partitions	Acoust cons. approved reduced rating	($30,000)
IC10	Millwork	1&2	Additional counters in offices	Design Develop add	
IC11	Millwork	1&2	Substitute black epoxy in lieu of gray	What is lighting impact?	($56,985)
IC12	Millwork	1&2	Provide bookcases ext wall -Lauff&Cima Labs	Design Develop add	
Elevator					
EV01	Handicap Elevator	2	Reduce new cab size/capacity	confirm acceptance / price(formerly E03)	
HVAC					
H04	Steam	1	Provide heat recovery at DCM	BR+A to provide more info	$20,000
H06	Steam	1	Provide clean steam generator for humidity.	Included w/8th flr (see IC04)	---
H08	Distribution	1	One fancoil for each bay at north		
H09	Controls	1&2	Reduce thermostat zones		
Plumbing					
P04	Risers	1&2	Delete at south corridor side	Part of systems design	($180,000)
P05	Piping	1&2	Retain copper branch piping	Need info from BR+A on the scope	
P06	Lab grade water	1&2	Scale back DI/RO distribution	verify during DD	
P08	Risers	1&2	Zone to delete PRV's	BR+A reviewing BCC suggestion	
P10	Risers	1	Relocate CW,steam,condensate,rain leader	Will gain program space	$213,000
P11	Risers	1	Relocate CW, & rain leader piping	Will gain program space	$117,775
P12	Lab grade water	1&2	Look at alternate RO system equipment	MIT to verify performance criteria	
Electrical					
E03	Handicap Elevator	2	Reduce new cab size/capacity	Now an Elevator item (see EV01)	---
E05	Lighting	1&2	Standard pendant lights in corridor	EAI studying alternatives	($50,000)
E06	Lighting	1&2	Reduce gen lighting in labs, add task		($29,656)
E08	Distribution	1&2	Utilize alternate cable tray		
Equipment					
EQ01	Autoclave/Glasswash	1&2	Substitute Consolidated autoclave #SR24DVMC	Architect to review	($150,000)

Figure B
Scope log—potential cost impacts, pending.

Massachusetts Institute of Technology
Buildings 16 and 56 Renovations
Potential Cost Impacts
PRICING

Major System	Item of Work	Phase	Description	Comments	Estimate
Structural					
S02	Foundation wall	2	Increase capacity of wall to support bridge	Geotech input required	
Exterior Construction					
EC01	Elevator	2	Alternate to glass block	Curtainwall is proposed	
EC02	Receiving Dock	2	Enlarge dock and enclosure	BCC will price Dock and vestibule scheme	
EC03	Vestibules	2	Reduce size of west vestibule	To be priced as part of dock	
EC04	Vestibules	1&2	Research alternate door configurations	EAI to look at alternatives	
Interior Construction					
IC06	Ceilings	1&2	Add Ceiling Fins @ corridors 2-8	Alt to clgs deleted per accepted IC05	
IC07	Laboratory casework	1&2	Refurbish casework for support rooms vs new	Kewanee budget vs new -($125/lf)	T B D
IC09	Millwork	1	Chngs in mllwk in lect hall & clsrms / blackout shades		
Plumbing					
P02	Sprinklers	1&2	Reduce shaft sprinklering	Now a Fire Protection item (see FP02)	---
P03	Sprinklers	1&2	Fire main conn to Bldg 68 pump	Now a Fire Protection item (see FP03)	---
Fire Protection					
FP01	Sprinklers	1	Revised design w/ NFPA criteria		
FP01	Sprinklers	2	Revised design w/ NFPA criteria		
FP01	Sprinklers	3	Revised design w/ NFPA criteria		
Electrical					
E07	Emergency Power	2	Increase Emergency Generator to 800 kw		
E12	Distribution	1	Refeed existing animal lab elect panels	Now inc w/ 8th fl pricing (see IC04)	
Equipment					
EQ05	Autoclave / Glasswas	1&2	Stacking units recommend by Architect	EAI to provide information	
EQ06	Laboratory casework	1&2	Refurbish casework for support rooms vs new	Kewanee-($125/lf)(formerly IC07)	T B D

Figure C
Scope log—potential cost impacts, pricing.

Massachusetts Institute of Technology
Buildings 16 and 56 Renovations
Potential Cost Impacts
REJECTED

Major System	Item of Work	Phase	Description	Comments	Estimate
Interior Construction					
IC 13	Floors	1&2	Provide seamless epoxy	VCT is the standard	$255,388
IC14	Code	1&2	Maintain bldg. separations	Not required by code	$52,160
IC15	Partitions	1&2	Provide veneer plaster at corr. 2-8	Gypsum wall board accepted	$36,383
IC16	Ceilings	1&2	Provide ACT ceilings at office & reception 2-8	Painted exposed structural clgs.	$50,337
IC17	Partitions	1&2	Provide 1 hr. rated wall construct. around elev. lobbies	Not required by code	
IC18	Program	2	Deduct link construction	Will include link	($634,647)
Plumbing					
P12	Distribution	1&2	Meter building 16 and 56 separately	Not required	$405,648
P13	Distribution	1&2	Delete gas shut offs at teaching labs	MIT requirement	$1800/EACH
HVAC					
H07	Exhaust	1&2	Provide high rise stair pressurization and smoke exh.	Not required by Code	$274,335
H10	Distribution	1&2	Combine radiation loops.		
H11	Distribution	1&2	Sheetmetal duct hangers.		
H12	Controls	1&2	Pneumatic vs. DDC at offices	All DDC	($16,000)
H13	Controls	1&2	Fume hood two position sash in lieu of VAV	Look at ganging in plenum instead	($250,000)
H14	Exhaust	1&2	Alternate exhaust duct material	Will carry as alternate	
H15	Exhaust	1&2	Reduce exhaust to BSC	Maintain exhaust standards.	
Electrical					
E02	Handicap Elevator	2	8 Stop electric elevator w/ machine rm.		$67,720
E11	Distribution	1	Maintain vertical busways	Will replace all new.	($18,052)
E12	Emergency Power	2	Temporary power from central MIT plant		($34,814)

Figure D
Scope log—potential cost impacts, rejected.

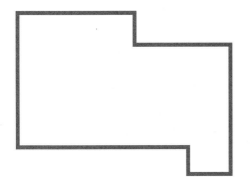

Index